Routledge Revivals

Science and Ethics

First published in 1942 (second impression 1944), this book forms
a debate about the endeavour to find an intellectual basis for ethics
in science.

Science and Ethics

First published in 1942, second impression 1943, this book makes a case about the relations of biology and the ethical basis for social structure.

Science and Ethics: An Essay

C. H. Waddington

Routledge
Taylor & Francis Group

First published in 1942
Second impression 1944
by George Allen and Unwin LTD

This edition first published in 2016 by Routledge
2 Park Square, Milton Park, Abingdon, Oxon, OX14 4RN
and by Routledge
711 Third Avenue, New York, NY 10017

Routledge is an imprint of the Taylor & Francis Group, an informa business

A Library of Congress record exists under LC control number: 43002080

ISBN 13: 978-1-138-95694-0 (hbk)
ISBN 13: 978-1-315-66542-9 (ebk)
ISBN 13: 978-1-138-95695-7 (pbk)

SCIENCE AND ETHICS

an essay by

C. H. WADDINGTON, SC.D.
Fellow of Christ's College, Cambridge

together with a discussion between the author and
The Right Rev. E. W. Barnes, Bishop of Birmingham
The Very Rev. W. R. Matthews, K.C.V.O., Dean of St. Paul's
Professor W. G. de Burgh, F.B.A.
Professor C. E. M. Joad
Professor Susan Stebbing
Professor A. D. Ritchie
Professor H. J. Fleure, F.R.S.
Professor J. S. Huxley, F.R.S.
Professor J. B. S. Haldane, F.R.S.
Dr. C. D. Darlington, F.R.S.
Dr. J. Needham, F.R.S.
Professor H. Dingle
Dr. G. Burniston Brown
Dr. Karin Stephen
Miss Melanie Klein
Miss Miriam Rothschild
Professor J. D. Bernal, F.R.S.
Professor Chauncey D. Leake

London
GEORGE ALLEN & UNWIN LTD

FIRST PUBLISHED IN 1942
SECOND IMPRESSION 1944

THE PAPER AND BINDING OF THIS
BOOK CONFORM TO THE AUTHORISED
ECONOMY STANDARD

PRINTED IN GREAT BRITAIN
in 11-Point Baskerville Type
BY UNWIN BROTHERS LIMITED
WOKING

CONTENTS

INTRODUCTION

THIS book is in the form of a debate about a thesis. The original essay which gave rise to the discussion was intended as a fairly systematic, although extremely summary, exposition of a point of view which had been implied in a short popular booklet which I had recently published under the title of *The Scientific Attitude*. The essay was submitted to the editors of *Nature*, who invited a number of authorities to comment upon it. Other authors were moved to contribute to the discussion, which became too voluminous for the correspondence columns of a weekly journal. The whole debate, both that portion which has not yet appeared in print as well as the original public discussion, appears to constitute a valuable contribution to a subject the profound importance of which is becoming ever more generally recognized. In collecting it together, and recording it in a form more permanent than a private correspondence, every attempt has been made to edit it in such a way that it does not lose the essential character of a discussion, that of being an interchange of views. The age-long endeavour to find an intellectual basis for ethics is an enterprise of such importance, and of such difficulty, that any explorer of that country must always be glad to hear the voices of his fellow-travellers. "This," Wittgenstein once said to me, "is a terrible business—just terrible! You can at best stammer when you talk of it." This book is communal, perhaps even co-operative, stammering.

C. H. W.

CHAPTER I

THE RELATIONS BETWEEN SCIENCE AND ETHICS

By Dr. C. H. Waddington

THROUGHOUT most of history, man's concept of the Good has been rightly considered to have, or at any rate to require, a philosophical justification; that is to say, a justification dependent on the characteristics, not of a particular individual, or group of indivudals, but of the world in general. This might be deduced from observation, as in the theory of Utilitarianism, or revealed by the voice of God or of conscience. During the last quarter of a century, four lines of thought have converged in an attack on this notion, and their combined effect has apparently gone far, at least among what may be called 'popular intellectual' circles, to rob ethical statements of any claims to intellectual validity. All four of these trains of thought had their origin in scientific movements. They were:

(1) The psycho-analytical, based on an examination of individual psychology, which seemed to imply that man's ethical system is a mere product of his early sexual reactions to family life, and has no more generality than that has.

(2) The anthropological, based on a comparative study of social systems, which tended to show that ethical beliefs differ extremely from culture to culture and can therefore have no general validity.

(3) The Marxist, primarily based on a study of the changing society of Western Europe, which appeared to assert that ethical systems are expressions of class forces and are epiphenomena which may be left out of account when we are considering the mechanism of social development.

(4) The anti-metaphysical of the Logical Positivists, based

on the attempt to realize the 'unity of science' through a study of meaning, and issuing in the view that ethical statements have no meaning of a verifiable nature.

None of these summary statements of the four arguments is, I think, an entirely fair account of the contribution which the science in question has made to the study of ethics. But they do represent not too inadequately the sense in which these contributions have been understood among wide circles of the general reading public, including many of the younger men of science. Taken together, the four lines of attack were undoubtedly successful in persuading many people that science either has nothing to do with the formulation of ethical systems, or even is necessarily inimical to any such attempt. I wish to argue here the contrary thesis: That if these four contributions are correctly interpreted, ethical judgments are statements of the same kind—having, as the logisticians would say, the same grammatical structure —as scientific statements. I shall deny Carnap's argument that the typical ethical statement 'killing is evil' is merely a paraphrase of the command 'do not kill',[1] and "does not assert anything, and cannot be proved or disproved". I shall argue that an ethical judgment is better typified by a statement such as "You are an animal of such a kind that you must consume 7 mgm. of vitamin C per diem, and should consume 100 mgm.", that is to say, by a statement which has scientific significance.

An ethical belief must be believed by someone; and the psycho-analytical discoveries, which are concerned with the development of the ethical systems of individuals, are the most profitable basis from which to begin an examination of the scientific basis of ethics. Psycho-analytical literature is voluminous, and is couched in a somewhat anthropomorphic jargon which, while it may be an inevitable result of attempting to write in conscious language of mental processes which do not occur within consciousness, is undoubtedly not very perspicuous for the layman. But one

---

[1] Carnap, R., *Philosophy and Logical Syntax*, Kegan Paul (1935), 24.

may, with all due diffidence, mention two points which seem to emerge from it.

In the first place, ethics appears among psycho-analytical phenomena as the consciously formulated part of a much larger system of compulsions and prohibitions. Many of these remain permanently below the level of consciousness, but, all together, they make up a more or less isolable dynamic function within the personality, known as the super-ego. By setting up the super-ego as the entity for investigation, psycho-analysts are abolishing, in a very radical way, the class distinctions which we commonly make among our inner compulsions, which lead us to hold that the prohibition on picking one's nose in public, for example, although often much stronger than that on lying, is less worthy of consideration. This is a piece of realism for which one can have nothing but gratitude. Moreover, it brings out clearly the very important point that one cannot avoid ethics; it is impossible to give them up like smoking in Lent. They are part of the super-ego, and the super-ego is inescapably among those present (accompanied by the ego, the id, the ghosts of Œdipus, Narcissus and the rest) whenever we do anything.

The second of the psycho-analytical results which requires attention is more fundamental, but in some ways less straightforward. Put shortly and crudely, it is that the super-ego is formed as a result of experience of the material world, and that its propositional content has been verified in experience. There are two difficulties in the way of establishing this. First, the super-ego is being formed from the age of about six months onwards, and empirical observation at that time has a peculiar character which it later loses. "The baby", writes Joan Riviere, "cannot distinguish between me and not-me; his sensations are his world, *the* world to him."[1] The first crude notion of externality, of otherness, arises through the experience of an inability to control; and the objects which thus intrude into the baby's solipsistic day-dream are inevitably personalized, distinguished as

[1] Riviere, J., *Love, Hate and Reparation*, Hogarth Press (1937), 9.

"not-me but another person". More than that, they must appear to butt in from outside what had been thought of as all-embracing. It is, I suggest, because the development of ethics is connected with this break-up of solipsism that it has that character of other-worldliness, of absoluteness, which made plausible the anti-metaphysical comment that one can no more talk about it than about the ultimate reality behind the world's appearances. "Wovon man nicht sprechen kann, darüber muss man schweigen", said Wittgenstein in 1919, addressing philosophers.[1] His words would have been more apposite in the mouth of a mother talking to her child; but unfortunately one screams as though the devil were on one's tail; probably he is.

The second difficulty in establishing the dependence of the super-ego on experience arises in connexion with the distinction between the external and the internal, between the individual and his environment. There is first a simple confusion to clear out of the way. One finds, for example, the following sentence by Freud:[2] "Whereas the ego is essentially the representative of the external world, of reality, the super-ego stands in contrast to it as the representative of the internal world. . . ." But the context makes it quite clear that Freud is speaking here of the adult personality, at a time when the super-ego has already been formed. He is not, in calling that entity the representative at that time of the internal world, denying that at an earlier period, during its formation, it was dependent on the external world. In fact, in another place he states, fairly explicitly, the point which I wish to make: "The role which the super-ego undertakes later in life is at first played by an external power, parental authority. . . . This objective anxiety is the forerunner of the later moral anxiety."[3]

[1] Wittgenstein, L., *Tractatus Logico-Philosophicus*, Kegan Paul (1919), concluding sentence.

[2] Freud, S., *The Ego and the Id*. Cf. 'General Selection from the Works of Sigmund Freud,' Hogarth Press (1937), 259.

[3] Freud, S., *New Introductory Lectures on Psycho-analysis*, Hogarth Press (1933), 84.

But the difficulty goes deeper than this. The author who has, perhaps, contributed most profoundly to our knowledge of the formation of the super-ego is Melanie Klein. Her view "lays emphasis on the importance of the impulses of the individual himself as a factor in the origin of his super-ego and on the fact that his super-ego is not identical with his real objects".[1] But, she writes, "In thus regarding the impulses of the individual as the fundamental factor in the formation of the super-ego we do not deny the importance of the objects themselves for this process, but we view it in a different light." Now it may be pointed out that in emphasizing the importance of the external objects in the formation of the super-ego, the role of the innate impulses of the individual has not been denied. The question at issue is whether the ethical beliefs which form part of the super-ego are injected into the individual apart from and independently of his experience of the material world, or whether they are formed by the interaction of the personality and the world; there cannot be any question of the super-ego being impressed by external circumstances on to a merely receptive and featureless individual. The answer which I am urging is that the situation is actually parallel to that with which we are familiar in genetics; all characters are, as Goodrich put it, both inherited and acquired; they are products of the interaction between the genes, which we usually consider internal, and the equally necessary factors, such as oxygen, nourishment, etc., which we usually consider external. Strictly speaking, one cannot say that the propositions of ethics arise from experience of external, as opposed to internal, connexions; their origin is the observation that the world is such, and the personality is such, that the individual must follow certain rules.

Here, it may be urged, the word "must" in the last sentence may be going too far. Granted that the propositions of ethics are derived from experience, does that experience

---

[1] Klein, M., *The Psycho-analysis of Children*, Hogarth Press (1932), 195, 197.

teach us more than techniques which lead to pleasurable results, and do we still need to invoke some non-experimental criterion to judge, not what gives us pleasure, but what *is* pleasurable or good and what bad? But if there were any such ulterior criterion, it would have to be of the most general and unspecific character. What we are considering is not the abstract entity 'ethics', but actual super-egos as they are effective in human personalities; and they are so variable from person to person, that, if their contents are taken to consist of rules for obtaining some ultimate objective, that objective must be of an extremely vague character. Further, there are many propositions for which it is clear that no ulterior criterion of value is necessary. The statement that it is as well not to put your hand in the fire is not based on anything else except the fact that if you do it will cease to be a hand: and existence is its own justification; hands are the kind of things which do not go in fires. Self-destruction of an entity only comes into question when there also exists some larger unit of which that entity is a part, and it only occurs when this more inclusive unit is more powerfully energized in the dynamic system of the super-ego.

According to some psycho-analysts, an urge towards self-destruction is, in actual fact, very early awoken in the young child. But there is obviously in existence an entity in which the child is only a part, namely society, and the facts which the child is learning and incorporating into his super-ego are very largely facts about the existence of society and his place in it. He discovers, for example, that if, in anger at being denied the maternal breast, he attempts to attack his mother, he is either restrained or at least disapproved of. That disapproval is ultimately based on nothing more than the existence of society, which would be impossible if aggression were uncontrolled. The child, of course, does not himself discover that the existence of the society of which he will be a member demands the control of *aggression*; that knowledge can only belong to his parents, and may not be formulated even in them. But the disapproval which the child experiences is a result, mediated either by intelligent knowledge or by the unconscious processes of natural selection, of the requirements of human society. The ethical principle 'Be good, sweet child!' derives what validity it has from social facts as real as the calorie quota for human survival.

During the very early months, when the main structure of the super-ego is being formed, the most important facts which come to the notice of the child are social facts, arising from its relations with its parents, nurse, etc. The anthropological discovery that systems of ethics differ in different cultures is therefore not only not surprising, but is indeed a necessary consequence, and a confirmation, of the view here put forward. The way in which these systems of social behaviour are conditions for the existence of the cultures concerned has been fully discussed by Malinowski and his followers. But we must, I think, go farther than this. Ethics, at this point in the argument, appears as a system of rules of action derived from the necessary conditions for the existence of society. They appear, that is to say, as simply conservative. It would be a sanguine man who would deprecate such a function at the present day, but we cannot in fact expect society to continue unaltered. A tendency to evolutionary or developmental change is a general characteristic of biological entities, including societies, and it is certainly true of Western European civilization that the ethical systems engendered within it are not simply conservative but are among the agents of this change.

The contribution which theoretical Marxism made to the study of ethics was actually not to debase ethics to the position of a mere epiphenomenon, but was a combination of this point with the anthropological argument mentioned above. The widespread misunderstanding of this is partly due to the very diverse, and sometimes regrettable, practical applications of the Marxist theses on ethics which have been made by various political parties; and partly to a certain naughty-boyishness, a roguish delight in paradox *pour épater les bourgeois*, in the Grand Old Men themselves. Such a spirit is perhaps not unexpected in professional revolutionaries, but it has led to some remarkable confusions when interpreted by the more earnest of the true believers.

Marx and Engels urged, first, that ethical ideas are derived from the experience of social facts. This part of their argument is one of the almost innumerable meanings of the famous phrase 'freedom is the knowledge of necessity', an epigrammatic statement the highly complex ambiguity of which should commend it to the school of poetic criticism represented by Mr. Empson. Further, they asserted that different social classes, encountering different material conditions, form different ethical systems. They also showed that the differing conditions of the social classes bring about developmental changes of the society as a whole. Since they, of course, acknowledge the fact that "all the driving forces of the actions of any individual person must pass through his brain, and transform themselves into motives of his will in order to set him into action",[1] this implies that it is only through the systems of beliefs to which they give rise that the social conditions are effective. The point was somewhat obscured by their insistence on what was the newest and most controversial aspect of their doctrine, namely, that the social facts from which the ethical systems are derived could be ultimately reduced entirely to matters of economics. And it was, as mentioned above, also concealed by some of their more irresponsible utterances; for example, by Engels: "it is precisely the wicked passions of man—greed and lust for power—which, since the emergence of class antagonisms, serve as levers of historical development",[2] in which he emphasizes the imperativeness of the socially determined Good by comparing it to unrestrained biological drives. But, in spite of the confusion caused by such verbal tricks, Marxism did provide the logical basis for the view

[1] Engels, F., *Feuerbach*, Lawrence, n.d., 62.   [2] *Ibid.*, 47.

that realist ethics can change society and not merely preserve it.

Having now reached the position of seeing a social system as something the existence of which essentially involves motion along an evolutionary path, we are confronted again with the question which was discussed five paragraphs above in terms of static existence: Do we need some external criterion to decide what is the 'good' direction of evolution, or is that implicit in the society? Again, I think, one can answer that no criterion external to the natural world is required. An existence which is essentially evolutionary is itself the justification for an evolution towards a more comprehensive existence; a society implies a direction of development into a society which could include the earlier stage, as, to take an exaggerated example, American culture can include that of the Red Indian, but not vice versa. One can put the same thing in another way by reference to the history of evolution; on the whole, the later products of animal evolution have capacities which include and transcend those of their ancestors.

But, it may be said, granted that the existence of a society does imply a direction of change, why should that direction be accepted as good? One could quote eminent authority against such a view. "Let us understand, once for all," wrote T. H. Huxley,[1] "that the ethical progress of society depends, not on imitating the cosmic process, still less in running away from it, but in combating it." But he was writing under the spell of that extraordinary impulsion, so incomprehensible to us to-day, which forced the Victorians to transmute the simple mathematics of their major contribution to theoretical biology into a battle-ground for their sadism. To Huxley, the cosmic process was summed up in its method; and its method was "the gladiatorial theory of existence" in which "the strongest, the most self-assertive tend to tread down the weaker", it demanded "ruthless self-assertion", the "thrusting aside, or treading down of all

[1] Huxley, T. H., *Evolution and Ethics*, Macmillan (1894), 83.

B

competitors". To us that method is one which, among animals, turns on the actuarial expectation of female off-spring from different female individuals, a concept as unemotional as a definite integral; and we can recognize that quite other, though equally natural, methods of evolution may occur when it is societies and not individuals which are in question. Moreover, being no longer hypnotized by the methods of evolution, we can see its results; and they cannot be adequately summarized as an increase in bloodiness, fierceness and self-assertion.

Huxley, in fact, was morally outraged by what he took to be the character of the cosmic process, and was therefore forced to exhort civilization to combat it. With our present ideas, the general character of the cosmic process, or as we should now say, of the course of evolution, does not seem so morally offensive that we cannot accept it. To return to our question, we must accept the direction of evolution as good simply because it *is* good according to any realist definition of that concept. We defined ethical principles as actual psychological compulsions derived from the experience of the nature of society; we stated that the nature of society is such that, in general, it develops in a certain direction; then the ethical principles which mediate the motion in that direction are in fact those adopted by that society. Of course the good is, as the anthropologists pointed out, different in different societies, and particular cultures which regress may be actuated by principles at variance with the cosmic process. But in the world as a whole, the real good cannot be other than that which has been effective, namely that which is exemplified in the course of evolution. It should be noted that this, if you will, cosmic fatalism, does not imply a fatalistic attitude to the evolution of any particular section of the world, for example, of the society of which one happens to be a member.

It is, then, finally clear that science is in a position to make a contribution to ethics, since ethics is based on facts of the kind with which science deals. And the nature of science's contribution is also clear; it is the revelation of the nature, the character and direction of the evolutionary process in the world as a whole, and the elucidation of the consequences, in relation to that direction, of various courses of human action.

But the practical difficulty remains. The fundamental features of an ethical system are formed, as part of the super-ego, in the very early years of life. A child learns at its mother's knee that aggression must be controlled; and it learns a very little later that taunting its younger brother's weakness is a form of aggression; but when does it learn that adopting an unscientific attitude to the social problem of nutrition is also aggression? Most of the scientific contributions to ethical thought are of a kind which seem, at the present time, difficult to convey in the early formative years in which the most effective features of the super-ego are being laid down. Perhaps this appearance is deceptive, and perhaps after a few generations the fundamental notions of the scientific outlook will be so deeply incorporated into normal life that they can be transmitted by the unconscious gestures of mothers and nurses. An adequate psycho-analytical study of people who have grown up in Soviet nurseries might tell us whether this is too wildly optimistic. But in any event we should do well not to neglect the second line of attack, but should study deeply how the intellectual content of the super-ego may be modified in later life, and the data which we can provide about the nature of the cosmic process appropriately attached to the powerful general principles about love and aggression which are by that time already in existence. It is the profoundest of scientific principles that a theory must work in practice; and that applies to scientific ethics no less than to the latest modification of the quantum theory.

CHAPTER 2

COMMENTS

1. By the Right Rev. E. W. Barnes, f.r.s.
Bishop of Birmingham

I FIND myself in fundamental agreement with Dr. Waddington, though I should base my argument on an epistemology
more explicit than his own. To start off, I would aver, with
Mach, that "bodies or things are compendious mental
symbols for groups of sensations—symbols that do not exist
outside of thought". The basis of all knowledge is experience.
So-called external objects are constructs from experience:
equally the doctrine of evolution and the view of the universe
summed up in the Ten Commandments are constructs from
experience. Of course, the experience may be partial:
elements in it may be false (that is to say, unconfirmed by
the majority of our fellow-men). The activity of the mind
which links together elementary perceptions and fashions
the constructed symbol may be inadequate to make a
symbol which shall cohere with other symbols as we try
to picture some wide region of the universe in which we
find ourselves. But by a process of trial and error, in which
the individual constantly checks his experience by that of
others, the race has gradually created, among other ideas,
those which we distinguish as external objects, laws of
Nature and ethical principles.

We assume that there is an external world of objects to
which our bodies belong. But, if that world exists, is our
picture of it correct? We cannot say, for we cannot transcend human limitations. Are our scientific laws accurate?
Probably not: they correspond, however, to humanity's
present state of mental development. Can we say that our
ethical standards and the commands by which we seek to

make them effective are sound? They, too, are as partial, as transitory, as our supposed knowledge of the spiritual character of the universe.

Are then our scientific laws and our ethical principles of no value? By no means. They are approximations to truth, nearer than those which were reached in the past and later modified or even discarded by the growing wisdom of the race.

Unfortunately, the problem of the mind-body relation is so intractable that it is difficult to say how far intellectual and ethical tendencies are inherited. I would agree with Dr. Waddington in affirming Goodrich's conclusion that all characters are both inherited and acquired. The genes carry certain modes of reaction to environment. A relatively homogeneous community is built of the same stock of genes changed to some extent by recurring mutations; and an individual born into it assimilates with especial ease the community's intellectual, social and ethical formulation of experience.

Is Dr. Waddington quite fair in his strictures of T. H. Huxley? The evolutionary process on earth, until the rise of the placental mammals with their increasing parental affection, was non-moral. "Nature red in tooth and claw" is an actual fact. Huxley was right in asserting that between man and the cosmic process as it has been, there ought to be war. The strongest objection to ethical theism lies in the fact that the creative process has been non-moral. But just as evolution has been a creative process in that new things, and in particular man himself, have emerged in it, so it may well be that the process itself is being transformed: no longer, it may be, are new animal forms being evolved, but new levels of spiritual understanding are emerging. Boutroux died twenty years ago, but his "Religion and Science" in *Contemporary Philosophy* is not out of date. He said: "According to the results of science herself, there is nothing to guarantee the absolute stability of even the most general laws that man has been able to discover. Nature evolves, perhaps even fundamentally." He added that, if the remotest principles of things are thus transformed, that very transformation must obey laws which are analogous to the immediately observable laws of experiment Are we wrong behind such change to find purposive activity, to postulate God as its source, and to see in the ethical change which results from the growth of human experience His progressive revelation of Himself?

2. By the Very Rev. W. R. Matthews, k.c.v.o.
Dean of St. Paul's

Comment on Dr. Waddington's important and interesting paper is difficult because it raises so many questions which are highly controversial. Only a treatise could deal with them all. I must confine myself to some rather disconnected jottings. Frankly, I am not quite clear about the main thesis. If it is that the natural sciences have a valuable contribution to make to the study of ethics, few would deny it; if it is, as I think, the contention that the central problem for ethics can be solved by the method of natural science, that seems to me a disastrous error. No doubt science can throw light on the way in which minds come to apprehend values but, as it seems to me, it cannot determine whether they are truly values or only appear to be such, nor can it determine the scale of values, if any.

A certain scepticism about some of the alleged findings of science may be permitted. For example, the super-ego appears to me to be a piece of useful mythology; probably it helps to "explain" the process by which we reach ethical maturity, but may it not be misleading to treat it as an "entity"? The important fact is that mature and sane men have ideals which, as they believe, commend themselves to their reason, and sometimes they have imaginary pictures of themselves as they know they ought to be. Again, the diversity of moral codes at different levels of civilization can be exaggerated. Virtues which are honoured among us, such as courage or even kindness, are honoured in crude and more limited forms by people of lower cultures. The development of moral ideas is not determined wholly by social condition; there is a dialectical development of the ideas themselves, and if it is true to say that societies create ideas, it is even more true to say that ideas create societies.

The use made of the psychological concept of "compulsions" perplexes me. As I understand it, a compulsion is an irrational and perhaps irresistible tendency arising from the unconscious. The moral experience in its authentic form is surely the opposite of a compulsion. The agent believes himself to have the responsibility of choice and the ethical "ought" is recognized not as something which must be obeyed but something which deserves to be obeyed, though it may be difficult and unpleasant. "Had it power (compulsion) as it has authority, it would absolutely rule the world." I am even more perplexed by what seems to be asserted about the goodness of evolution or even of all existence. "We must accept the direction of evolution as good simply because it *is* good." I think I must have failed to grasp this point, because in the preceding sentence we are told that revised ideas about evolution enable us to feel that it is not morally offensive, as T. H. Huxley thought it was. This seems to imply that Dr. Waddington has considered the course of evolution and found that it is not morally offensive. Now, how, on his own principle, could he possibly do that? What criterion did he apply? No doubt, as a theist I am bound to hold that there is a direction in evolution or rather that organic evolution is a part, perhaps a very small part, of the Divine purpose, but I see no reason to suppose that at any given moment the actual direction of evolution is towards higher values, and this is pre-eminently the case when the process is largely determined by human will.

There is a most fundamental problem raised for ethics by the evolutionary hypothesis. I wish that Dr. Waddington

had said more about it. Shortly it is this: evolution appears
to suggest that all moral ideas are relative, but the moral
consciousness regards some of them as absolute and unless
it does so the moral life is simply abolished. We are confronted with the situation now in every home. There are
some things of such value that men ought to be prepared
to die for them; it is reasonable to be prepared to die for
them. Why? Men answer with action and, it may be sus-
pected, deplorably confused notions of ethical theory; but
they act because, in their simple way, they believe that the
voice of duty comes from a Source deeper and more intimate
than the course of evolution.

3. By Professor de Burgh, f.b.a.

I join issue with Dr. Waddington on two points. First, when
he offers, as a typical example of a judgment that is at once
ethical and scientific, the statement "You are an animal
of such a kind that you must consume 7 mgm. of vitamin C
per diem, and should consume 100 mgm.". I see nothing
ethical here at all. The rules acquire ethical significance
only when in a given case I judge the effort after survival,
to which it prescribes the means, to be morally right or
wrong. If I am the father of a family and there is only
a limited supply of vitamin C available, it may be my moral
duty to throw the rule to the winds and forego the means
to my survival. The 'must' of the rule is not the unconditional 'ought' of morality, but the condition of attaining
an end, as to the morality of which the rule says nothing.
The 'should' in the last clause is ambiguous; it may mean
either 'you ought to' or merely 'you will have a better chance
of surviving if you do'. The former meaning alone is ethical,
but I fancy that Dr. Waddington intends the latter. He may
reply that he sees no difference between the two, any more
than when on a later page he identifies what is pleasurable
or what leads to pleasurable results (two different matters,
by the way) with what is good. We seem to be back in the

dear old days of Herbert Spencer. Do fallacies never die, however often they are confuted? If 'you ought' is identical with 'you'd jolly well better', and if 'this is good' is only another way of saying 'I find this pleasant', then the moral consciousness is an illusion and a cheat, and the sooner we stop talking about it the better.

Dr. Waddington puzzles me, again, when he argues that the evolutionary process itself supplies us with a criterion of good, and that we need no other. I fail to see what he means by saying that this "cosmic fatalism does not imply a fatalistic attitude in the evolution of any particular section of the world", for example, of one's existing society. The 'psychological compulsions' with which he identifies ethical principles are surely, in his view, determinant of every act of every citizen in every race and age. If so, morals, whose business it is precisely to draw 'class-distinctions' among our natural impulses, vanish from the picture. Moreover, what ethical criterion can be derived from the scientific doctrine of evolution? Biology knows nothing of the qualitative distinction of higher and lower, better and worse; it can only display the continuity in the modifications of species through descent, showing what form of life succeeds what, and that certain more complex organisms have less complex organisms as their temporal antecedents. If the second law of thermodynamics should work its will and if all mind and all life should be eliminated from our planet, the process would be just as much an evolutionary process, in the sense relevant to biology, as that by which man has arisen from the ape. Apart from ethical presuppositions read in from other and non-scientific sources, evolution has no concern with value. The cosmic process is not indeed, as Huxley thought, immoral, save for those who indulge the 'pathetic fallacy' and interpret it in the light of their own emotions; but it is wholly amoral. The scientific study of it cannot teach us what is good or what we ought to do. It cannot even say 'must' in its predictions; it can tell us only what has been, what is, and what, in varying measure of probability, will be in the time to come. It cannot tell us that what will be is right or good.

These are my two grounds of dissent from Dr. Waddington, and I think they are fundamental to the issue. With much else in his article I cordially agree. But I venture to add a remark that travels a little beyond the scope of his discussion. It seems to me important to grasp the bearings of this amorality of Nature on our present world troubles. Are they not in large measure due to the fact that our knowledge of science, especially in its practical applications, has outrun, far outrun, our morality? Science has placed instruments of world-shaking power in the hands of rulers who abuse them for their own unrighteous ends. These instruments are in themselves, like physical Nature, non-moral. Neither Nature nor science is to blame for their misuse by man. Morality lies in the will to good, immorality in the will to evil, that is, in the choice of ends, not in the means to their attainment. Of those ends, whether they be good or whether they be evil, science, for all its glory, can tell us nothing.

4. By Professor C. E. M. Joad

I propose to touch very briefly on those points in Dr. Waddington's article with which I agree, although, even where I agree, I cannot resist the temptation of entering a disclaimer against his uncritical taking over lock, stock and barrel of the pretentious jargon with which psycho-analysts disguise the commonplaceness of their observations upon the obvious. What, for example, does all this talk about the super-ego and its imposition upon the personality—is it, for example, upon "a merely receptive and featureless individual" or upon one who is "himself a factor in the origin of his super-ego"?—really amount to? That there is an individual person exhibiting certain specific characteristics which distinguish him from others—my dislike, for example, of the taste of marzipan, or my delight in the smell of privet; that this individual is born and grows up in an environment and that his resultant beliefs, including his ethical beliefs, are the result of the impact of the environment upon the characteristics which distinguish him from others, as well as upon those which he shares with others. That, as it seems to me, is all that Dr. Waddington and Melanie Klein are saying, and, put like that, it scarcely seems to justify the fuss.

I agree again with Dr. Waddington's interpretation of Marxism. I agree, that is to say, that Marx *did* provide for changes in, as well as conservation of systems of, social ethics, while retaining my private opinion that the real agents of ethical change are to be found less in the factors that Marx and Dr. Waddington emphasize than in the appearance of an ethical 'sport' in the shape of a Christ, a Buddha, a Socrates or a Blake who points the way to new levels of conduct and new standards of value to which in course of time the accepted moral codes of society as a whole gradually creep up. Or don't creep up! If they don't, then, to adopt a biological metaphor, the 'sport' has failed to breed true. I deliberately employ the biological metaphor in witness to my belief that the process of evolution still proceeds by 'mutation', although the scene of its operations has now been largely transferred from the physical to the mental and spiritual spheres.

So much having been said by way of not very impressive agreement, I come to my two major quarrels.

About the first I must say very little, not because it is not important but because it is subsidiary to Dr. Waddington's main thesis. He says that, if the contents of super-egos are taken to consist of general rules, they must be rules "of an extremely vague character". In more familiar language, the deliverances of men's moral consciousnesses vary so much that no general ethical principles as to what is good and right can be laid down.

I deny it, and claim that we do in fact all know, and always have known, that unselfishness is better than selfishness, kindness than cruelty. What is more, we can all recognize a case of cruelty when we see it and know that we ought to try and stop it—(the fact that we usually do not try is not to the point). I should go further and maintain that we do all of us know the sort of way in which we ought to live; that we know, in fact, that we ought to live very much as Christ enjoined. We may say that Christ's prescription for good living is wholly impracticable or is much too difficult; but that does not alter our conviction that it is the right prescription. The difficulty about ethics is not that we don't know what is right and know with a good deal of particularity, but that we lack the will or the ability to act in accordance with our knowledge.

Secondly, on Dr. Waddington's main point, I cannot understand how anything can be measured without a ruler which is external to and other than what it measures. Now to adjudge a movement as good or as bad—witness in this connexion Dr. Waddington's talk about "the 'good' direction of evolution"—entails that some meaning is understood to be conveyed by the words good and bad which serves as a standard of measurement by reference to which the movement is evaluated. Now this meaning cannot itself be part of the process which it is invoked to evaluate, any more than a ruler can be part of the length which it measures, or a man can lift himself by his own braces. Dr. Waddington points out that later stages of evolutionary development include the earlier. Certainly they do, but what of it? The later stages of a travelling snowball include the earlier, but that does not mean that the snowball's journey is ethically valuable or worthy of praise. It may not even be well advised; if it is heading for a precipice it is ill advised. The point is surely obvious enough. When Dr. Waddington affirms that evolution is moving in the right direction or is progressive—it is "good", he says, "simply because it is· good"—he is applying ethical standards to it. Now all progress implies movement in a direction and direction implies a goal. If I put myself in the Strand and set my legs in motion, there is movement or process, but until I know whether I want to go to Charing Cross or Temple

Bar I cannot say whether I am progressing or not. But the goal cannot be part of the process which seeks to realize it.

Once this is understood, it will be seen that the kind of question which Dr. Waddington is putting, when he applies the notion of 'right direction' to evolution and then proceeds to inquire whether our present direction is "right", is, *if we are to proceed on his premises*, like the question "Is it better to take the right fork or the left?" when asked by somebody who does not know where he wants to go; while further questions relating to the speed of the advance are like asking whether it is better to travel in a 40- or a 10-h.p. car, when you don't know where you are travelling, or whether it is good to travel at all.

5. By Professor Susan Stebbing

In commenting upon Dr. Waddington's article, the need to be brief compels me to concentrate upon a single point and to say too shortly what requires to be argued with the help of detailed examples. The point I select for comment is that the contribution of science to ethics lies in its revelation of "the character and direction of the evolutionary process in the world as a whole", and that the examination of this direction will yield the criterion of human action. Although I am in agreement with much that Dr. Waddington says here and in his little book, *The Scientific Attitude*, I find a serious difficulty in understanding his present argument. He maintains that the "real good" is that which has been effective, that is, that which has been exemplified in the course of evolution; accordingly, he argues that "we must accept the direction of evolution as good simply because it *is* good according to any realist definition of that concept". Presumably the word "must" in this sentence means "are logically compelled", so that our acceptance is an admission of what follows logically from the "realist definition" of good.

It is not, however, clear whether this is what Dr. Wad-

dington means since he at once proceeds to drag in the
notion of fatalism, in order to ward off a possible charge
of being fatalistic. But such a charge would not make sense
if I have correctly interpreted the phrase "we must accept".
The difficulty is increased when we take note of the context
in which the sentence I have quoted occurs. Dr. Waddington
is disagreeing with T. H. Huxley's protest against accepting
the cosmic process as the standard of ethical progress. The
answer he makes consists of three parts, or—as I prefer to
put it—he gives three different answers: (1) the method of
evolution is to us—as contrasted with Huxley—"as unemotional as a definite integral"; (2) the results of evolution cannot be adequately summarized as an increase in
bloodiness, etc.; (3) the course of evolution does not seem
to us now "so morally offensive that we cannot accept it".
But (3) seems to me to make a muddle of the argument.
If good is defined as that which is effective, that is, that
which is in the direction of evolution, what is the point
of answer (2)? And if the concept upon which the method
of evolution turns is unemotional, then why, again, bring
in (2)? In short, it is not compatible with Dr. Waddington's
"realist definition" of "good" to speak of the course of
evolution as morally offensive or morally admirable. But
his answer (2) suggests that he does think it necessary to
show that Huxley was mistaken in his estimate of the blood-
thirsty character of the struggle for existence. Suppose
Huxley's estimate had been correct: would it make sense
to say that the evolutionary process was morally offensive?

6. By Professor A. D. Ritchie

I have read with great interest Dr. Waddington's lucid and
well-reasoned essay in speculative metaphysics, into which
he has ingeniously woven hypotheses derived from Freud
and Marx, but I fail to see the alleged connexion between
science and ethics. He says that the contribution of science
to ethics is "the revelation of the nature of the character

and direction of the evolutionary process in the world as a whole, and the elucidation of the consequences, in relation to that direction, of various courses of human action". (This might almost be a quotation from Herbert Spencer.) The direction of the evolutionary process may have been revealed to Spencer or Dr Waddington, but not by science. It is said that Amœba and Hydra represent early stages in animal evolution, yet there are plenty of them alive still. For all we know they may survive long after Homo has perished by mutual slaughter. Would that make them better or worse from the scientific point of view?

The process of evolution has thrown up Hitler, Himmler, Goebbels and their like. If they were to win the War, would that show the direction of the evolutionary process? Evolution has produced the nightingale and the kingfisher we admire; also Sacculina, the parasite of the common shore crab, and also the matrimonial habits of spiders, which we do not admire. Does science tell us which is better? I select these examples because they are of no evident economic importance and our judgments may be considered disinterested. I am not arguing that these judgments of approval or disapproval are subjective or irrational, only that they are outside the scope of science. By reason of its method the only values within its scope are truth and error as judged by logical consistency and conformity to fact. If the logical positivists confined themselves to this assertion they would be on safe ground. I am not arguing, either that Dr. Waddington's theory is wrong, only that, like every ethical theory (including the theory that there are no ethical distinctions or that they are meaningless), it rests on *a priori* presuppositions it is best to be honest about.

On a minor point, I must protest against the notion that it is a recent discovery that different societies have different moral codes. It seems to have been known to the author of the *Odyssey*, and certainly to Herodotus a few centuries later. Lastly, may I recommend Dr. Waddington (and others interested in the relations of science and ethics) to read *Five*

Types of Ethical Theory by Professor C. D. Broad, where he will find his own type of theory labelled and docketted; and specially to read p. 284—the last page but one?

7. BY PROFESSOR H. S. FLEURE, F.R.S.

Much appreciation is no doubt widely felt for Dr. Waddington's statement that, if various modern theses are correctly interpreted, ethical judgments are allowed by them to be "statements of the same kind as scientific statements". One also agrees with his view that the putting forward of these theses has somehow persuaded many people of a lack of any link between science and ethical systems. This seems a natural temporary reaction belonging to what Samuel Alexander called the deanthropizing phase of thought. For millennia, men have sought authority for social codes in anthropomorphs created by their imagination outside the evolutionary sequence and empowered to insert into it new items—dispensations they have been called—from time to time. The comparative method in the study of man, outstandingly represented by Frazer, has vividly suggested that what were held to be impregnable rock-fortresses of traditional belief are, rather, erratics in the moraines of folk-lore. The old authority has gone. It withered too, at a time when an individualist age was obsessed with the idea of Nature red in tooth and claw, and even a Huxley could suggest that men's ethical systems must stand in antagonism to the cosmic process.

In their various ways Alexander, Lloyd Morgan, Smuts and Sherrington are trying to get us beyond the inevitable phase of disorientation. Unlike older systems, the work of science must not claim to give us something complete and unchanging; it must have ever-recurring readjustment as its keynote. Would that those who are busy making blueprints of a better world would realize this; so many of their schemes are static! Perhaps a main contribution of the humanist at the present juncture is the thought that man is a social being, and that, within society, there is an unceasing and not always successful struggle towards freedom of conscience, towards replacement of external by internal controls. One may add that the survival-value of this freedom is related to the facts of observation and inference, namely that life's history on earth has been a process of ever-recurring readjustments, and that, with few exceptions, the fate of those forms which did not readjust has been extinction. At the same time, it should be remembered that these developmental adjustments are selective; if some features are enhanced, others are atrophied. So it is not very wise to suggest that the later include the earlier; that unduly simplifies the idea of change and suggests acceptance of the rather crude notion of the inevitability of progress.

8. BY PROFESSOR J. S. HUXLEY, F.R.S.

Out of the breakdown of traditional systems of thought, glimmers of new light appear, islands of solid land emerge out of the chaotic flood. Dropping metaphor, the question is whether any new system of thought, sufficiently strong to provide the foundation for living, can be evolved in time to substitute reintegration for disintegration. As science has played a major part in bringing about the disintegration of the old, it should attempt to do at least as much in the new integration.

Dr. Waddington's interesting article is a valuable contribution to this. As he points out, psychology, anthropology and sociology have largely contributed to the breakdown of traditional views on ethics. He might have added many other sciences. Evolutionary biology is one, with all its implications as to human ancestry, the struggle for existence, and the abolition of the idea of purpose in evolution. All the physical sciences have contributed, by providing a mechanistic explanation of natural phenomena previously attributed to supernatural powers and often invested with an ethical aura—witness the legend of the rainbow in the

Old Testament, or the frequent view of lightning, floods or earthquakes as expressions of Divine anger. Similarly, physiology and pathology have removed deformity and infectious disease from the ethical sphere; they are no longer considered as Divine retribution for moral lapses.

When it comes to the constructive side, I have little to add to Dr. Waddington's interesting thesis. He might, I think, have pointed out that in some cases science indicates a new ethic, or at least a new type of ethical approach to old problems. This may be illustrated by my last example. We can no longer believe that pestilence has any connexion with moral lapses in the conventional sense, or with the failure to observe certain rituals or to believe certain dogmas; but we can lay down certain new types of moral duty arising out of the nature of infection—duties both individual and social, concerning cleanliness and the prevention of disease and of its spread.

I have two specific comments. One concerns the basis for the quality of absoluteness and otherworldliness possessed by the super-ego and the systems of ethics for which it is the vehicle. Dr. Waddington makes what I believe to be the quite novel suggestion that this is connected with the breakdown of the solipsistic early phase of the child's existence. While this may be a contributory cause of the otherworldliness, I cannot feel that it accounts for the absoluteness, for the fact that certain aspects of morality are felt as a categorical imperative. The origin of this, as I have elsewhere suggested, must more probably be sought in the all-or-nothing method adopted in higher animals for avoiding conflict. This has been proved to operate to prevent conflict between antagonistic muscles and between competing reflexes. Observation shows that it must also normally apply to competing instincts in sub-human vertebrates. Finally, all we know of human psychology indicates the strong probability that it operates in repression in early life. Man is the only organism in which conflict is normal and habitual, so that some form for minimizing its effects is essential; and this will be of the greatest importance in early childhood, before sufficient experience has been accumulated to enable conflict to be dealt with empirically and rationally.

The antagonistic forces which hold down repressed ideas and impulses are kept away from the main body of consciousness; hence the apparent externality of ethical law. They are held there by the strong but automatic processes of repression; hence the compulsiveness of the super-ego. And repression is, or attempts to be, total, seeking to keep certain impulses wholly out of consciousness; hence the all-or-nothing character of the ethical prohibitions of the super-ego.

Some repressions are more complete than others; and in many cases the degree and method of repression can be modified or the prohibitions of the super-ego transferred in their operations from one field to another. Hence we may say that a great part of our ethical development will consist in diminishing the absoluteness and compulsiveness of our early categorical imperatives, and in altering the field to which they apply, in the light of reason and experience.

Put in another way, we may say that primitive and absolutist ethics, based on the non-rational and unconscious processes of the mind, inevitably tend to limit human activity by locking up conflicting psychological 'energies' in the repressive mechanism of the unconscious. For constructive and truly humanistic ethics, we need to liberate these forces from their unconscious grappling, through reason and still more by appropriate education and by opportunities for fuller living.

The other point which I would like to make is perhaps even more fundamental. Dr. Waddington writes: "an existence which is essentially evolutionary is itself the justification for an evolution towards a more comprehensive existence." While this is true, it is so general as to smack of Panglossic optimism. It is an observed fact that the majority of evolutionary trends are either irrelevant to progressive change, or are even opposed to it in direction, or are inherently limited specializations. As I have set out at some length elsewhere (in the first essay in my book *The Uniqueness of Man*) evolutionary progress can be objectively defined, and further is a rare phenomenon; the potentialities of further true progress now appear to be restricted to our own species, though there is no guarantee that we shall achieve them. The problem here is thus to study the possible directions of change; to decide which make for progress and which do not; which make for unlimited and which for limited progress; and to attempt to adjust our social systems and our ethical ideas in such a way that, as Dr. Waddington rightly points out is possible, they should form a mutually reinforcing whole, making for the maximum speed of progress in the correct direction.

Dr. Waddington points out the difficulties arising from the fact that the ethical systems of different societies differ enormously, one conception of the good often contradicting another. Here again there is an evolutionary parallel. Thanks to the work of Sewall Wright, we know that small and isolated animal and plant species will often show 'accidental' differentiation, which is not necessarily biologically advantageous, and may sometimes even be disadvantageous. The same appears to apply to the evolution of cultures.

Further, as Darlington has pointed out in his recent book, *The Evolution of Genetic Systems*, certain evolutionary changes may be of immediate advantage, but of eventual disadvantage in robbing the stock of evolutionary plasticity and adaptability. Here again there are doubtless parallels from ethics. The short-term efficiency of ruthless State dictatorship as opposed to the inevitable long-term triumph of more humanistic systems is a case in point.

With such modifications, Dr. Waddington's thesis of ethical systems as indispensable social organs, derived from the impact of a changing external world on the minds of individuals via the social environment, but themselves then helping to effect changes in the external world and the social environment, appears to be a fundamental one, and worthy of the most careful study.

9. A Reply by Dr. C. H. Waddington

I would like to make a reply to some of the very interesting comments on my essay on the relations of science and ethics. I will try to be brief, although some of the questions raised are probably due to the brevity of the original essay. I will not allude to all the points of agreement, which are actually more numerous than some of the writers suggested. Thus I entirely agree with Dr. Matthews that ideas affect societies; in fact I supported this point by a quotation from Engels.

There are two major issues: whether ethical principles are founded on our experience, and the problem of free will. In the former connexion the arguments of my opponents have been stated by Professor Joad in a form which is so nearly a *reductio ad absurdum* that much of my work has been done for me. "I cannot understand", he writes, "how anything can be measured without a ruler which is external to and other than what it measures." By this he certainly does not mean merely that no system of mensuration is possible with less than two objects; for after all I am not suggesting that an ethical system involving different degrees of good would be engendered by a universe consisting of a single indivisible act. Nor, I presume, does he refer to the fact that our units of measurement, though roughly specified by the nature of the world, are in detail defined arbitrarily. His remark only provides a basis for his subsequent argument if it is taken to mean that we determine the relative sizes of objects by reference to some transcendental foot-rule reached down from beyond the boundaries of space and time. This, as we know from the theory of relativity, is untrue. The space-time framework is a function of the material objects lying within it. I might indeed have expressed my main contention by saying that, just as space-time issues from the material world, so the ethical system could be logically derived from our experiences. I may assure the Dean of St. Paul's that I am not urging that we should immediately reject all ethical principles which we cannot in practice trace back to biological and sociological data any more than I suggest that we should all learn enough mathematics to convince Sir Arthur Eddington that we fully understand the logical structure of an inch. I am merely concerned to show that the validity of ethical principles can be accepted even if we reject any criterion imported from outside the perceptual universe.

Joad again reduces to absurdity the view that there is an ethical criterion independent of experience by his statement that we all know (innately is implied by his argument) that "we ought to live very much as Christ enjoined". There are a thousand localities, from Dachau to Dahomey, where it is impossible to assert this with any plausibility. We prefer the ethical intuitions of Christ, Buddha or Socrates to those of Hitler or Rosenberg not because they are more mystical but because they seem more likely to carry society forward in the direction it has already taken. I see no grounds for rejecting the view which I put forward on the basis of psychological and anthropological evidence, that our tendencies towards sympathetic behaviour, although of sufficient strength to have enabled man to develop a degree of social existence, are nevertheless merely one of the general drives towards various unspecific forms of behaviour by which his conduct is affected.

The argument given above must serve, in the space available, as a reply to Professor de Burgh, who demands an unconditional validity for my example of an ethical scientific statement, but is apparently willing to forgo it for the contrasted 'Thou shalt not kill', the ethical nature of which he would not deny.

The widespread disagreement with my argument about evolution is a continuation of the same dispute. In the first place, I am glad to find that the course of evolution has been revealed, not only to myself and Herbert Spencer, as Professor Ritchie suggested, but also to Professor Huxley; surely the attempt to repudiate the normally accepted evolutionary sequence on the grounds that certain primitive animals still exist is a forlorn crusade. But the crux is my statement that the direction of evolution is good simply because it *is* good. By that I meant that if the ethical system is to be derived from the nature of the experimental world, we must pay attention to what that world is like; and one of the most important data is the scientifically ascertained course of evolution. The remark about fatalism which followed was intended to indicate that the general trend may suffer temporary set-backs. Just as loss of structure due to parasitism, as in Sacculina, does occur but is not typical of the greater part of evolutionary change, so social regressions, of a spatially or temporarily limited nature, are easily conceivable. This point was expanded by Professor Huxley, with whose remarks both here and in his valuable essay "The Uniqueness of Man" I am in substantial agreement.

Professor Stebbing attacks an outlying bastion of my position in this field; not my actual discussion of evolution so much as my comments on T. H. Huxley's remarks about it. My "three different answers" to Huxley's argument were answers to three different questions. The first was a rejection of his description of its *methods*, the second of its *results*. In the third I countered the possible objection that, logical though my derivation of the good might at first sight appear to be, I had actually identified it with something which no one had ever dreamed of calling by that name. I was arguing that my conception was not only self-consistent but also not unrelated to the conventional meaning of "good".

A more difficult point is raised by those who suggest the possibility of a general and persistent regression, for example by the operation of the classical second law of thermodynamics. I think that the difficulty, if it should at any future time actually arise, could only be got over by a theory of levels of ethics. One could distinguish a social good, dependent on a good derived from human individual biology, which again would be dependent on the effective principles of change in the physico-chemical world; and one would have to be content to deduce that a continual increase in physical good gives rise to an undulation in the development of biological good. But difficult though the problem undoubtedly is, that difficulty does not arise only on my theory of ethics; how can it be surmounted if ethical values are attributed to a beneficent deity?

The other fundamental disagreement, which relates to the problem of free will, was raised explicitly by the Dean of St. Paul's and Professor de Burgh, and implicitly by several others. I confess myself unable to offer a satisfactory reconciliation of materialistic determinism and the efficacy of the human will; but again the problem is not one for my theory alone, and I shall be agreeably surprised if my commentators are in a much more comfortable logical position. I can only make one suggestion, with the greatest tentativeness. First, I suggest that it may be more profitable, in discussing this matter, to picture the human mind not as a simple mechanism of stimulus and response, but as containing a set of drives (each, figuratively speaking, a complicated motor) one or other of which can be set in motion by pressing the appropriate switch. In the decision whether, say, the sex or the nutritive drive becomes activated, the external stimuli not only reach directly for the switches, but also bring into play internal systems whose functions are also to affect the choice of which drive is selected. It may be that the sensation of an effort of will is no more than the conscious symptom of the activity of one of these internal systems, perhaps the super-ego or some part of it. If this part of the mind is, owing to the way in which it has been derived from the external world, normally effective in the direction taken by the evolutionary process as a whole, then could it not be argued that its conscious correlate is in fact indicative of a good impulse, quite independently of whether strict causation is violated or not? The peristaltic action of our bowels, although not under much conscious control, is good on my definition; may not the sensation of a deed well done be just as valid, on a higher plane, as the sensation of physical well-being after exercise and a bath?

It may be helpful at this point to suggest an analogy between my treatment of man's ethical behaviour and the normal scientific treatment of his feeding behaviour. Different men may have almost any notion of what constitutes good food; notions derived from social suggestion or accidental circumstances during their upbringing. In attempting to improve nutrition, we might begin by defining food in some other terms, e.g. as that which satisfies the pangs of hunger, just as Utilitarians define the good as that which gives pleasure. We see now that the latter definition is inadequate, and the former is not in fact the kind of thing which science attempts to formulate. It proceeds, instead, by investigating the function of food. The function can be objectively described; and it is described in terms of the carrying forward of an observed process, in this case the process of normal growth and development. It is a complicated matter to describe what is normal, as opposed to abnormal, growth, but it can be done; and once it is done, there is a generally valid criterion of goodness in food, which can be used to persuade the milk-drinking nomad of the Steppe, the rice-eating Chinaman, the white-bread addicts of England, to improve their diets. Similarly I suggest that the psychological and Marxist investigation of the ethical systems of individual men show that their function is to implement an observed process, of progressive evolution; and if that is so, we have a criterion of universally valid ethics from which the individual variants can be criticised.

CHAPTER 3

FURTHER COMMENTS

1. BY PROFESSOR J. B. S. HALDANE, F.R.S.

DR. WADDINGTON's interesting essay suggests that he is still wavering between the theory that when you have explained a thing you have explained it away, and the fundamental but usually unspoken postulate of science that everything has an explanation, even though this implies an infinite series of causation. In Dr. Waddington's opinion, Marxists say that ethical systems are epiphenomena which may be left out of account when we are considering the mechanism of social development. T. H. Huxley invented the word 'epiphenomenon' to mean a mental event caused by physical events, but not in its turn causing physical events. I believe that in the long run science has no room for such loose ends. Certainly Marxism has not. "It would be totally absurd", wrote Lenin in *Materialism and Empirio-criticism*, "that materialism should maintain the 'lesser' reality of consciousness." Marxists hold that mind is real, but secondary to matter because matter existed before mind. Similarly, they think that economic and social structure largely determines the ethical system. "Thou shalt not commit adultery" is meaningless in a society with no marriage. "Thou shalt not steal" is replaced by "Thou shalt not waste" as property becomes socialized.

Given such a point of view, ethics must be fitted into our world picture, though we cannot yet see how in full detail. It is clear that ethical practices and ideas have a history, both in the development of communities and of individuals. Marxists have stressed the former process, Freudians the latter. This fact does not mean that ethics are arbitrary or baseless. England is real enough though it was once under the sea; vision is real though human embryos have no eyes. It does mean that we cannot act as rightly as possible without a study of contemporary history, which shows us what is alive and growing and what is vestigial in current ethical systems. Perhaps the careful attempts to isolate university staffs from the impact of history, if they have been advantageous to abstract speculation, have disqualified them from valid judgments on the highly concrete problems of right and wrong. The technique of modern warfare, which has broken down this isolation, may lead them to more realistic thinking on ethics.

A fuller study of the literature of Marxism might, I think, not only have shown Dr. Waddington that Engels stressed the importance of unconscious motivation before Freud, but also have made him more sympathetic to T. H. Huxley's thesis in *Evolution and Ethics*. Stated in dialectical terms, it is that the cosmic process, which was responsible for human evolution, negates itself by generating the ethical process. The problem then arises of how man is to continue evolving if the congenitally weak are not killed off. Hitler's solution is substantially to abolish ethics. The correct solution will not be so simple. There is a real contradiction, which will be resolved when men not only realize, as eugenists do, that they ought to control their own evolution, but also possess, as they do not at present, the knowledge and technique necessary for this control.

2. BY DR. C. D. DARLINGTON, F.R.S.

The earlier contributions to this discussion of science and ethics seem to show by their extreme diversity how far thinking men may still be from an understanding of the scope and method of science.

Science is concerned with what a man (or a thing) must do, ethics with what he thinks he should do. Until the contrary is proved, therefore, we must suppose ethics to be derivable from science. How it is to be derived is a

question on which scientific men cannot yet be expected
to agree. But its historical relationship with the subject-
matter of science, with the material conditions of society,
is surely a commonplace. The one undergoes evolution,
so does the other; and by evolution we mean the irreversible
succession of changes which seem to be characteristic of
all integrated systems. Take the Christian ethic. It has
suffered three major recensions on its journey from the
Sea of Galilee to the City of London, each of them well
suited to the social and political conditions in which it
has in fact proved fit to survive. Meanwhile other systems
have arisen outside our own by revolution, and proceeded
afterwards by an evolution faster than any we have known.
As Professor Julian Huxley has suggested, some of these
systems may have sacrificed the credit of permanent adaptability for the cash of immediate advantage. That is a
question which events are now deciding. In doing so these
events have already displayed the somewhat Hobbesian
principle which would be too obvious to repeat if it were
not so often avoided, namely, that individuals will serve
the State in proportion as they believe that the State will
serve them. This is true at any given moment with little
regard to whether that State's government is inherited,
elected, or imposed, or all three together.

How the State is to serve the individual most efficiently
will therefore depend, under rapidly changing conditions,
on the adaptability even more of its ethical than of its
political system; and our conditions are changing very
rapidly. Evolution is no longer a hypothesis. It is happening
on our doorstep. Now in all evolution there is a lag in the
adaptation of one part of the organism to changes in
another. The more extensive and more highly integrated
the organism, the greater the lag. Our society is both extensive and highly integrated, and even the horrors of the
industrial revolution in the North scarcely disturbed an
ethical system emanating from the comfortable South.
Now, however, the situation is different. We are faced, as

before, with changing internal conditions. But our system is also in conflict with a second divergent system and at the same time in combination with a third, equally divergent.

It might be supposed that such a crisis is more likely to be resolved by empirical action than by analytical thought. That would be a mistake. Already during the present War political expediency has led to violent changes in the relations of the individual to society, changes which scientific method could have directed long ago and without any such compulsion. Why then should we not prepare for any worse emergencies by applying scientific method before they arise?

It may be objected that these questions for which a scientific solution is offered are not ethical but political. On the contrary, they are both ethical and political: the distinction lapses as soon as both are subjected to scientific treatment. What a man must do and what a man should do are always the same for the man himself at the moment he does it. In such measure as men submit to scientific discipline that sameness becomes extensible to the whole commonwealth. For universal agreement at successive levels of analysis is not merely the aim of science. It is an aim which experience shows has always been attained.

Science is therefore bound to be the foundation of the ethics of the future and of a system of ethics with some expectation of that universality which has hitherto failed mankind.

3. By Dr. J. Needham, f.r.s.

It was an excellent idea to base a general discussion on the relations between science and ethics on Dr. Waddington's stimulating and lucid account of the subject. What has been most striking about the comments which have been made on it is the failure which some of the commentators exhibit to understand his view of the nature of the evolutionary process. The persistent existence of the lowest forms of life (to which Professor Ritchie directed attention), or the fact that parasites may achieve a high degree of adaptation to environment at the cost of profound degeneration, or the continuation of evolution (in Professor R. A. Fisher's phrase) "in the teeth of a storm of adverse mutations", have nothing to do with the inescapable fact that, during biological evolution, the degree of complexity and organization has increased. With the appearance of man, the maker and user of tools, the speaker, the moulder of his surroundings, this process, the outward and visible sign of which has been a progressively greater independence of the organism *vis-à-vis* its environment, reached its culmination. Thinkers such as Herbert Spencer (whom some of the contributors go out of their way to attack), were perfectly correct in viewing social evolution as continuous with biological evolution. In social evolution we cannot but see a more or less continuous rise in level of organization parallel with the increasing size and complexity of human communities, culminating in the conception of the world co-operative commonwealth now dawning upon the minds of men. Though there have been backslidings innumerable, there have also been points higher than the main curve of human social evolution sweeping its way across the graph of history.

Some of the contributors seem to be still under the influence of the Darwinian preconception which saw nothing in animal life but the struggle for existence, a concept which, as Engels carefully pointed out, had been introduced from Malthus's analysis of the predatory characteristics of capitalistic society. But there were others beside Spencer who showed the onesidedness of the idea of Nature red in tooth and claw. Kropotkin pointed to the very value of animal associations in this struggle, and Henry Drummond (a much misunderstood thinker) successfully traced the beginnings of social altruism downwards to the numerous phenomena of parental care and even to the donation of part of the self for the succeeding generation in every reproductive act. Drummond even went so far as to say that the goal of evolution was love and the good life, an assertion which his biographer described as "grotesque", but which we can scarcely think so if we recognize, as we must, the highest levels of human co-operative social life as themselves the products of evolution.

This, I take it, is what Dr. Waddington means by saying that the evolutionary process itself supplies us with a criterion of the good. The good is that which contributes most to the social solidarity of organisms having the high degree of organization, which human beings do in fact have. The original sin which prevents us from living (in Professor Joad's phrase) "as Christ enjoined" is recognizable as the remnants in us of features suitable to lower levels of social organization, anti-social now. There is, of course, the incidental difficulty of continually modifying the letter of the teaching of the great ethical 'mutants' to fit changing techniques and increasing knowledge without losing their spirit.

From this point of view, the bonds of love and comradeship in human society are analogous to the various forces which hold particles together at the low colloidal, molecular, and even sub-atomic levels of organization. Henry Drummond actually dared to say this. If such an idea is accepted, Professor Joad's insistence that we must have some extra-natural criterion of ethical values ceases to have any point. The kind of behaviour which has furthered man's social evolution in the past can be seen very well by viewing human history; and the great ethical teachers, from Confucius onwards, have shown us, in general terms, how men may live together in harmony, employing their several talents to the general good. Perfect social order, the reign of justice and love, the *Regnum Dei* of the theologians, the Magnetic Mountain of the poets, is a long way in the future yet, but we know by now the main ethical principles which will help us to get there,

48 SCIENCE AND ETHICS

and we can dimly see how these have originated during social and biological evolution. Professor Stebbing is perplexed as to whether we ought to call evolution morally admirable or morally offensive; it is surely neither. The good is a category which does not emerge until the human level is reached.

For the benefit of Professor Ritchie, I may add that whatever label or docket in Professor Broad's book is attached to the views here expressed is a matter of relative indifference to me. They certainly cannot be called original. Many others have appreciated the emergence of ethical relationships and their interpretation in the light of scientific thought.

4. BY PROFESSOR H. DINGLE

A scientific statement is essentially an expression of relations derived from and applicable to experience: it is therefore easy to determine whether a statement is scientific or not by considering its relation to experience. Dr. Waddington's statement "The real good cannot be other than that which has been effective, namely, that which is exemplified in the course of evolution", is clearly not derived from experience, for it does not express anything found by observation. Nor is it applicable to experience; when we try to apply it to any actual ethical problem (for example, "Is it morally good to bomb German cities?") it is found to be useless.

I do not believe that Dr. Waddington intends to be among the apriorists, but actually his so-called scientific ethical principle belongs to the company of Eddington's inviolable laws and Milne's cosmological principle. It provides one more example of the widespread abandonment of science in the name of science.

5. BY DR. BURNISTON BROWN

I am sorry that Dr. Waddington allows the word 'good' to be spelt with a capital, even if only once. The use of a capital letter makes an adjective appear to be a noun, that is, a thing, which has an independent existence, and this leads to endless confusion, such as that involving 'Eternal Values', etc.

A more serious lapse (especially from one who has written on the scientific attitude) is the lack of definition of the terms used. Clear definition is essential to the progress of science, for the facts upon which its theories are built cannot be verified unless they are expressed in clearly defined terms which enable other research workers to establish similar conditions for observation or experiment. Now when we consider the subject of ethics we find at once that the words 'good' and 'evil' have never been clearly defined, and consequently the application of scientific method is impossible. Words are, of course, only symbols, and unless we know clearly how they are related to events in our actual lives, that is, their meaning, the use of them in sentences is mere word-spinning and leads only to confusion.

As regards the intimate connexion between science and ethics, I should like to repeat, in a more pertinent form, a point of view which I put forward in an essay-review of the Bishop of Birmingham's Gifford Lectures (*Science Progress*, 116, 729; 1935):

(1) We strive for the greatest mental and bodily well-being, that is, happiness (fact of experience).

(2) This is greatest when others are also happy (fact of experience).

(3) To achieve (1) we should therefore strive for "the greatest happiness of the greatest number".

(4) To achieve (3) we require knowledge of facts about the actual world, and what would be the results, or probable results, of given actions in it.

(5) This knowledge is most reliably obtained by the exercise of scientific method.

(6) In order, therefore, to distinguish between good and bad conduct (good conduct being defined as that which conduces towards the greatest happiness of the greatest number and vice versa), we require knowledge obtained by science. Thus science is intimately connected with ethics.

Sections (1) to (6) might be said I think, to form a basis for a scientific ethic. This is not a static conception, for with the continual increase in our knowledge, it might happen that an act formerly thought conducive to the greatest happiness of the greatest number would be found not to be so. This is an advantage in a world in which the only certain thing we can say about the future is that it will be different from the past.

6. A REPLY BY DR. C. H. WADDINGTON

Professor Dingle has picked out of my essay a sentence which, given the definitions with which I was operating, is a tautologous expansion of the argument. He appears to have thought that it was intended as an empirical statement, and he denies that it actually is empirical. From this basis he proceeds to reject my opposition to the apriorist view of ethics on the grounds that the opposition is itself apriorist, since it is not based on observation. He even states that it has no application to experience, although it clearly implies that in making an ethical choice we should pay more attention to the probable effects of the alternative courses of action in relation to the scientifically ascertained direction of evolution than to our own or other people's ethical intuitions or any system of ethical rules, etc.

The whole misunderstanding depends on the implicit adoption by Dingle of the traditional, and to my mind quite unsatisfactory, theory of the nature of an ethical aim as something absolute and without history. Thus, in a recent publication,[1] he wrote: "It is clear that since the [ethical] principles of action must in essence be independent of the consequences of action, these latter being usually unknown, they cannot be expressed in terms of a rationalization of past experience". Now the grounds advanced here for the independence of principles from their consequences are quite inadequate, since the consequences of our actions are never certainly known even when we guide them by an obviously empirical working hypothesis. One suspects that the independence is asserted merely on the basis of the introspection of an adult man who disregards entirely his own development. But however it has been arrived at, this view discounts at the outset the possibility of observing the genesis of aims, and thus any statement about their origin must appear non-observational. The apriorist view in fact becomes a tautology, since it has been smuggled into the discussion at the very beginning under cover of a theory of nature of aims in general.

It is, however, by no means impossible to observe the genesis, and thus the nature, of an aim; I mentioned in particular psychological and anthropological observations. The possibility of such a study has been overlooked in traditional thought partly because of the late appearance of an interest in evolutionary and developmental problems in general, and partly on account of the spurious 'absoluteness' of ethical aims, towards an elucidation of which both Professor Huxley and I made suggestions. But it is the total neglect of such considerations which lies behind both the simple objections of Professor Joad and the more sophisticated ones of Professor Dingle. It also robs of much of their cogency the discussions of Professor Broad, to which Professor Ritchie referred me, couched as they are in terms of non-developmental concepts of 'reason', 'emotion', 'pleasure', etc.[2]

[1] Dingle, H., *J. Aristotelian Soc.*, 122 (1939).
[2] The discussion with Professor Dingle is carried on in greater detail in Chapter 5.

It is again an awareness of developmental considerations which I miss in Dr. Burniston Brown. The utilitarianism which he puts forward can, I think, be regarded as one of the historical forerunners of the line of thought which I suggested. His formulation, however, neglects all the advances made in our understanding in the last hundred years, and does not refute, or circumvent, the well-known difficulties of the theory. Thus he takes me to task for not defining my terms, and indeed the comments on my article show that in many cases I was not successful in indicating the subject of my remarks; but his first premise seems to me untrue if 'pleasure' is defined in the ordinary way, and without significance if the definition is adjusted to make the sentence true. I think these difficulties are surmounted, and that Professor Dingle's objection cannot be sustained, if, instead of saying that to achieve goodness we should strive for the greatest happiness of the greatest number, we state that our ideal of goodness is presented to us by a certain part of the personality, that the function of this part is the furtherance of evolutionary progress, and that the task of reason is to clarify that aim.

It may be emphasized here that I did not merely define the good as that which tends to promote the ultimate course of evolution. In science one does not, except when teaching mechanics in an old-fashioned way to third forms, define concepts in the sense in which geometrical concepts are defined and which allows deductions to be made. A scientific definition, which I hope was the kind I was employing, consists in indicating the phenomenon which one intends to call by a certain name. What I did was to use 'ethics' in the first place for the ethical judgments of an individual. I then advanced three propositions about such judgments; first that they are a part of the super-ego, secondly that they are built up as a result of experience; and thirdly, that the function of the super-ego is to implement those aspects of the personality (such as those on which social life depends) which are the most recently evolved. My statement that "the direction of evolution is good simply because it *is* the good according to any realist definition of that concept" is a summary of those three propositions; I am sorry to find that it is apparently such a deceptive summary, but perhaps the critics might be asked why they always omit the latter half of the sentence.

I am in general agreement with the remarks of Dr. Darlington and Professor Haldane, although the latter does not seem to have penetrated far enough into my essay to have discovered this. I also accept in general the thesis so ably argued by Dr. Needham, that the ethical principles formulated by Christ and the great ethical teachers are those which have in the past few thousand years tended towards the further evolution of mankind, and that they will continue to do so in the foreseeable future. In putting forward such a case it is easy to become involved in a circular argument. One can assume that the doctrines of the great ethical teachers are valid, and then assume that the evolution of mankind has been progressive because there has been some advance, however slight, towards those teachings. That is not sufficient for the thesis I am maintaining. My claim, and I think Dr. Needham's, is that the doctrines of the great ethical teachers have made possible the establishment of new types of social order which demonstrably allow of a fuller development of man's individual and social powers. For instance, the Christian ethic, by for the first time combining a deep respect for the individual with a low regard for relations of dominance and submission, released an enormous store of initiative for the arts of peace.

7. ADDITIONAL NOTES ADDED BY DR. G. BURNISTON BROWN TO HIS LETTER PRINTED ON PAGE 49 ABOVE.

(a) As regards the first premise above, the definition of greatest happiness as the greatest mental and bodily well-being of the individual, conforms with the requirements of clear definition (so far as it is possible in the subject of ethics) in that it indicates the nature of the experimental test required: a competent physician and a competent psychologist can estimate mental and bodily well-being with considerable accuracy—quite sufficient for this scientific ethic to be applied for the next thousand years!

(*b*) The point of view referred to in the above letter was summarized as follows:

"Let us consider the 'problem of dirt'. Chemistry and physics have shown that all matter consists of atoms which are neither 'clean' nor 'dirty': these terms are merely relative to our personal outlook. Similarly, biology and psychology have shown that all living matter exhibits the phenomena of self-preservation and reproduction, which are neither good nor bad: these terms are merely relative to our individual outlook. Thus an inoffensive atom combined with other equally inoffensive atoms can produce some of the most offensive odours by which the human olfactory system may be assailed: on the other hand, combined in a different way, it may produce the smell of a rose. Likewise the instincts mentioned can produce modifications which are extremely unpleasant, but equally they may produce others amongst the most highly praised by mankind. These considerations do not make cleanliness and uncleanliness any less real, but they do prevent us from idealizing them and saying that the universe can be explained causally by supposing that it exists for the attainment of perfect cleanliness. We still have to fight against dirt (chiefly because of bodily pain, i.e. disease), and scientific knowledge shows us how to set to work. Similarly, these considerations do not make goodness and evil any less real, but they do prevent us from idealizing them and saying that the universe can be explained causally by supposing that it exists for the attainment of perfect goodness. We still have to fight against evil (chiefly because of mental pain), and scientific knowledge can help us in our method in this case also. Notions as to what constitutes the criterion of cleanliness have varied from time to time and place to place, but the standard will probably continue to improve with the help of science measured, (say, by the decrease in world total of bodily pain). Ideas of what constitutes goodness have varied just as much, but may be expected to improve with the help of science (measured, say, by the decrease in world total of mental pain)."

CHAPTER 4

SOME PSYCHOLOGICAL CONSIDERATIONS

A CORRESPONDENCE WITH DR. KARIN STEPHEN

a. Comment from Dr. Stephen

Dr. Waddington has presented us with a most stimulating article which has provoked a discussion so fascinating (and tangled) that I should greatly like to be allowed to take my part in it. Let it be stated at the outset that I approach the fray from the side of psycho-analytical theory and its clinical findings. From this point of view Dr. Waddington's statement of his psycho-analytical case, though partly correct, does not appear to be entirely accurate, which is not to be wondered at considering the language in which, as he justly remarks, psycho-analytical exposition is often couched, and also in view of the extreme unfamiliarity of the subject-matter and the piece-meal fashion in which Freudian psychopathology has evolved itself. I should like to begin by summarizing Dr. Waddington's statement of the psycho-analytical case and then to annotate his statement with such emendations as seem to me to be required.

Dr. Waddington says: "Ethics appear among psycho-analytical phenomena as the consciously formulated part of a much larger system of compulsions and prohibitions . . . altogether they make up a more or less isolable function within the personality known as the Super-ego. By setting up the Super-ego as the entity for investigation the psycho-analysts are abolishing, in a very radical way, the class distinction which we commonly make among our inner compulsions . . . a piece of realism for which one can have nothing but gratitude. Moreover, it brings out clearly the very important point that one cannot avoid ethics . . . the Super-ego is inescapable . . . whenever we do anything."

Taking a minor point, to begin with, rather for the sake of elucidation than disagreement; according to the Freudian view compulsions are motivated, not only by the Super-ego, but equally also by the forces of the Id, i.e. the vital impulses, with which the Super-ego is dead-locked in conflict and which it is struggling to repress. Positive compulsions aiming at Id satisfactions occur when the repression is not completely successful and the forces of the Id succeed in breaking through and controlling thought and action, in spite of the Super-ego's desperate opposition, while on the other hand prohibitions or compulsions designed to frustrate the Id, more and more violent in proportion to the menace from the Id, and equally compulsive because equally repudiated by another part of the self (in this case by the Ego) emanate from the Super-ego. This, by the way, is, I think, the real explanation of the 'other-worldliness' of these compulsions and prohibitions which Dr. Waddington discusses later on. His suggestion that this alien quality is the result of the invasion of infantile solipsism by external reality is interesting (there is a lot more to be said in this connection about the gradual differentiation of self from not-self in the first few months or years of life which I must reluctantly leave on one side in this comment as it would take too long to go into it), but I believe that the feeling of 'other-worldliness' to which Dr. Waddington refers is produced essentially by dynamic forces inside the personality whose object is to isolate and boycott unacceptable impulses and hold them apart from the rest of the self. This work of repression against the Id is carried out in the unconscious by the primitive Super-ego. When it is not altogether successful there comes what Freud calls a 'return of the repressed' which is experienced as if it were a compulsive foreign influence, perhaps even, in extreme cases, as a demonic or divine invasion of the personality. Nothing ever feels so utterly alien as an impulse which really belongs to the self, but which is being repudiated in the unconscious.

The more radical criticism which I feel should be made

of Dr. Waddington's statement of the psycho-analytic theory of the Super-ego is his identification between 'good' or ethical, and whatever the Super-ego demands or prohibits. I do not think anyone would be prepared to maintain that everything that anyone may feel compelled to do is necessarily right, even if he regards the feeling of compulsion as coming from conscience. It is quite certain that the Super-ego is not always a reliable guide in matters of good and evil. The Super-ego is the force behind the still small voice, but any-one who has studied the vagaries of conscience, even in its more or less healthy and mature manifestations, must agree that this small voice (or raging dictator) though sometimes ethical in its demands, may, and at times does, inspire appalling behaviour whose results have been disastrous for humanity by any conceivable criterion of 'good'. And con-science, as we recognize it, is only a small and relatively highly ethical portion of the Freudian Super-ego. Its less normal behests in the form of neurotic or psychotic compul-sions may well land their victims in gaol or the madhouse.

Primitive or diseased Super-egos, when they are not out-grown, are the greatest danger with which humanity has to contend, worse even than tempest, flood or pestilence. According to Freudian psycho-pathology neurotic and prob-ably a large part of psychotic disease results from a dead-locking of the vital impulses of human beings with these ruthless and blind types of Super-ego.

This is perhaps the point at which I may best outline my own attempt to contribute constructively to Dr. Wadding-ton's discussion on Science and Ethics. I am warmly in agreement with Professor Huxley about the close connexion between 'evil' and what he calls 'the locking up of the "energies" by the repressive mechanisms of the unconscious' (these repressive mechanisms correspond broadly with what I mean by the dangerous ruthless blind Super-ego) and with his suggestion that the way towards 'good' is to be looked for along the lines of 'releasing these energies from their grapplings'. With his plan for doing this, e.g. 'through reason

and still more by appropriate education and by opportunities for fuller life' I am much in sympathy, though
reluctantly, I am bound to say that I doubt whether reason,
education and more opportunities would prove sufficient
by themselves, to achieve this liberation in all cases and
solve the problem of mental disease which is the result of
this locking up of energies. Reason and education (especially
if this be interpreted in a way quite other than its usually
accepted meaning of acquiring intellectual knowledge) and
opportunities for full living should undoubtedly be given
every possible chance, and it may be that they can carry
us a considerable way, how far it is impossible to forecast
as yet since the opportunities they have been given hitherto
have been too restricted to enable us to estimate their possible
benefits. At any rate I agree with Professor Huxley that
domination by a blind and autocratic Super-ego dead-
locking with the energies or vital impulses is destructive to
the human personality and that its modification of this
mechanistic type of self-regulation would produce 'good'.
My difference with him would come in only over the
practical question as to how this 'good' is to be promoted
and 'evil' to be reduced, and even here I should not disagree, but merely wish to supplement, because I fear he
may under-estimate the difficulties with which we are faced.
To attempt a full discussion of the causes and prevention
or cure of mental disease in such a comment as this would
be out of the question, and I dare not trespass on your
space to try to outline, even in the roughest way, the in-
credible complexity of the problems involved which seem to
centre round the breaking of an extremely tight vicious
circle. But it is a pleasure to find how closely Professor
Huxley's general attitude to Ethics coincides with the
approach to the question indicated by the findings of
Freudian psychopathology.

The theory underlying the view of 'good' and 'evil' to
which we both subscribe seems to be that the subject-matter
of Ethics is human personalities: 'evil' would coincide

roughly with neurosis and psychosis, i.e. with mental and moral disease, and 'good' with spiritual growth, health and sanity. This, in fact, is my own present working hypothesis with regard to ethical problems and I believe it is really very like Dr. Waddington's, since human personalities seem to be important among the products which evolution has achieved and so at which it may be presumed to have aimed. Moreover they seem, as I shall explain in more detail in a moment, to show a spontaneous tendency to evolve further in the direction which I am proposing to call 'good' and which I identify with mental and moral health. If we may identify the goal of evolution with this tendency then I believe the view I am putting forward as to the meaning of 'good' is exactly the same as Dr. Waddington's.

Working against this tendency, however, there appears to be a counter-tendency, which I call 'evil', towards mental and moral disease which arrests and corrupts this evolutionary process and I do not know whether we are justified in excluding this tendency from the total scheme of evolution. The progress made by this 'evil' tendency may be due to the institutions of our particular culture and thus may be remediable, provided human beings are not too mentally and morally ill to be willing and able to undertake the task of altering their own unhealthy culture. *Quem Deus perdere vult prius dementat*. I do not see at this stage, how we can know whether they are capable of this reformation or not, although we must act on the assumption that they can do it. But if both these conflicting tendencies must be regarded as equally included in the scheme of evolution then it seems to me that we must admit that evolution is divided against itself into two warring factions, one 'good' and one 'evil', and that we must suspend our judgment as to what the final outcome of the battle will be.

Returning now to Dr. Waddington's account of the Super-ego. He began by describing it as the basis of Ethics. I have explained that I consider this too sweeping. According to psycho-analysis, the Super-ego itself has an evolutionary history in the life of each individual. In its finally developed mature form its rulings may more or less coincide with what is good, but by then it will already practically have evolved itself away and been replaced by, or merged with, what psycho-analysis calls the Ego. In its earlier primitive forms it is a most unreliable ethical guide, partly because it is out of touch with reality and in its diseased form it is a terrible menace and the very reverse of 'good'. Dr. Waddington goes on to say 'The Super-ego (together with the Id and the Ego) are in everything we do.' This would probably be endorsed by psycho-analysis, but here it is a question of what kind of Super-ego, how primitive or diseased, at one end of the scale, or, at the other, how mature and how nearly merged with the Ego? As the Super-ego evolves and matures it changes radically. From the blind, automatic, all-or-none reflex, so ably described by Professor Huxley, deadlocking of the vital impulses of the Id, it changes, in the course of its normal healthy development, in the direction of a more flexible self-regulation, closely in touch with reality and guided by intelligence, hindering or promoting impulse discharge, in the light of its growing appreciation of and contact with external conditions, to further the aims of an equally evolved and matured Id, in so far as reality permits of this. As it approaches the end of its development the now transformed Super-ego is thus no longer at loggerheads with the Id, conflicting with it and stultifying it, but, on the contrary, falls more and more into harmonious alliance with it. This transformation which maturity produces in both Id and Super-ego, and which appears to be the goal of their evolution, brings both more and more closely into touch with reality, and so approximates them more and more to the Ego, whose function is to mediate between impulse-life and the external world so that satisfaction may become an actuality and the organism can function to its fullest capacity. (I must apologize for appearing perhaps extravagantly Utopian in the foregoing account of spiritual evolution. I am not so optimistic as to pretend that many, or even any human beings actually achieve this transformation of Super-ego into Ego, with the corresponding modifications in the Id which this must necessarily involve. I put this final consummation forward, however, as the ideal end-point of an already observable trend which may indeed be the goal (still extremely distant) of human evolution, and which, if it were ever attained, would solve most of our problems. In the meanwhile it at least supplies us with the definition of 'good' for which we have been looking.)

As matters stand at present we cannot pretend that anyone to-day ever achieves full ethical maturity, but it is already true that some personalities do seem to manage to carry development in that direction further than others, and in them certain significant modifications may be observed following a general direction which may be described as being away from the mechanical repetition of the reflex pattern and towards something different which, in contrast, we may as well call 'life', even though we may not, at present, be able to give a satisfactory definition of what we mean by 'life'. It may turn out to be no more than an improved edition of mechanism, or we may find that it is radically different. For the moment all we can say is that living behaviour does not seem to be quite the same as mechanical action and we can point to some respects in which the two differ. Living behaviour seems to be more flexible and adaptable, it even shows signs of learning, i.e. changing its mode of reaction in the light of experience, whereas mechanical action seems to be rigid. (No machine has ever been known to heal itself from injury, or to find a fresh way out of an unforeseen emergency!) We may even consider the possibility that life creates novelty, while mechanisms only repeat. Sometimes it is asserted that life is "free" in some sense in which, by contrast, machines are bound.

It would seem to be the sensible thing to put up, at present, with our inability to say the last word about life, and to make a careful study of the two opposed tendencies which appear to exist, towards living behaviour, in one direction, and mechanical activity, in the other. The development of the Super-ego seems to be a case in point. The behaviour motivated by the Super-ego appears, in the course of its evolution, gradually to change from the compulsive pattern to what may be described as a voluntary pattern springing from the harmonious co-operation of the various parts of the personality, Id, Super-ego and Ego, all tending to approximate more and more closely and to become merged in a unity which might be called the mature Ego: away from blind, fatal, driven repetition to "freedom", which is somehow connected with absence of inner friction and conflict, adaptation to reality and the creation of new behaviour patterns to meet fresh emergencies. This seems to me to be a tendency in the direction of life. (I recognize that my attempts at a definition of these two terms "life" and "freedom" are interdependent and so circular but I cannot at present do any better.)

The relevance of all this to our present discussion is that, although, if I were asked to offer any proof of my views, I should be unable to do so, I am suggesting that the tendency in human personalities in the direction of life is also a tendency in the direction of "good". There seems to be a spontaneous movement in this direction in human personalities, in so far as they are healthy, a tendency towards what I have called "maturity" which finds expression in relatively harmonious, frictionless, "free", voluntary, reality-adapted behaviour. Personalities progressing in this direction I propose to call "good". On the other hand mental disease seems to stunt, arrest, or even to reverse this tendency, and manifests itself in relatively blind, reflex, automatic, compulsive action. Personalities so afflicted I propose to call "bad".

If this be granted it does not follow that all "good" personalities will behave in accordance with one common standard and all "bad" ones in accordance with another.

Under different cultures the various personalities may conduct themselves in widely divergent ways, none of which could be regarded as right or wrong, in any absolute sense, but only as culturally correct or incorrect. The only universal judgment that could be made would be that, regardless of the cultural background and the particular type of behaviour which it enjoined, certain personalities (i.e. those which were internally harmonious and whose behaviour was regulated by their Egos) would be more or less absolutely 'good' and those who were racked by internal conflict and whose actions were compelled by their primitive Super-egos or rebellious Ids would be more or less absolutely 'bad'.

Such ethical judgments imply nothing in the way of praise or blame any more than praise or blame is involved in judgments such as we commonly make that, e.g., cancer is bad and a sufficiency of nutrition is good. I doubt however whether ethical judgments of this latter type are as fundamentally true as these other judgments about the goodness or badness of personalities which are here offered for consideration, since they seem to me to be derivative and to owe their ethical qualities rather to their effects on the development of personalities than to anything intrinsic in themselves. It seems to me that generally speaking it is easier for well-nourished, physically healthy personalities to carry through the evolution of their Super-egos and Ids in the direction of successful, harmonious maturity, and since whatever favours this will be 'good' as a means, on my definition of good, while whatever hinders it will be 'bad', it would follow that cancer and malnutrition are 'bad'. If, however, as some religious teachings suggest, pain, deprivation and illness favour good personality development, then far from being bad they will be good as a means of health and satisfaction.

And now, one last word about Dr. Waddington's account of what Freudians mean by the Super-ego. In discussing its origins he remarks: 'The Super-ego is formed as a result of experience of the material world . . .' [i.e. it tends to forbid behaviour which has led to painful experiences, e.g. disappointment, disapproval, punishment, loss, etc.]' 'The disapproval which the child experiences is the result . . . of the requirements of Society.' This was certainly a view at one time advanced by Freud when he first recognized the important part played in the formation of the Super-ego by the introjection of the parents, but Freudian theory has since developed considerably beyond this relatively simple view, and now regards the development of the Super-ego as considerably more complicated. Melanie Klein has made important contributions to the understanding of the early stages of Super-ego formation. Dr. Waddington quotes her as "laying emphasis on the importance of the impulses of the individual himself as a factor in the origin of the Super-ego—in fact that his Super-ego is not identical with his real objects". What Mrs. Klein means here is that the child's picture of its parents is actually very different from the originals, and that it is this picture which is introjected to form the basis of its own Super-ego.

But there is still a further point. Dr. Waddington quotes her as "regarding the impulses of the individual as the fundamental factor in the formation of the Super-ego". What she means here is *not* simply that the child has a nature of its own which reacts with the requirements of its parents and later of Society. What is meant is that the child projects its own unacceptable impulses on to the outside world and that it is these very same projected impulses of its own which it re-introjects and sets up inside itself as its Super-ego. This means that if, for instance, it experiences primitive impulses of rage or cruelty from which it takes flight because its Ego is too weak to manage them, it may be obliged to deal with them, instead, by externalizing them, by projection on to its parents, and thus it builds up a fantastic picture (called an Imago) of these outside people modelled on its own impulses. This picture will be cruel, murderous and also unmanageable and overpowerful like its own repudiated impulses, and when such an Imago is introjected to form the child's own Super-ego this will behave ruthlessly and cruelly to its unfortunate victim, just as savagely in fact as the child itself wanted to behave when it experienced the impulses which, in its panic, it was driven to project outside itself. The same vital energies which provided the driving force behind the child's own impulses of cruelty, revenge, murder, or whatever it may have been will now reanimate the reintrojected Imago which constitutes its tyrannical and cruel Super-ego. This point has been missed by Dr. Waddington and the misconception is carried on by Professor Joad in his contribution which shows no real understanding at all of the matter under discussion. It seems a pity, in view of the existing paper shortage, that Professor Joad, before rushing into print, did not trouble to find out the answer to his question 'What does all this talk about the Super-ego and its imposition upon the personality really mean?' If he would study the curious phenomenon of compulsive behaviour, most clearly exemplified in some types of obsessional neurotics, but, in its minor manifestations, extremely widespread, and would then familiarize himself with Freud's psychopathology, the crux of which centres round intra-psychic conflict, he would begin to get some inkling of the answer he is looking for (if indeed he really is looking and his question is not merely rhetorical); this study might however still take him some years, as his serious investigation of the matter does not seem yet to have begun.

b. Reply by Dr. Waddington

Dr. Stephen has amplified in many important ways my extremely summary statement of the psycho-analytical theory, but I do not find that her emendations controvert, or indeed entirely meet, my main thesis. I am of course aware that there are many psychological impulses not included in the super-ego; for instance, id impulses. But I think I am right in saying that our ethical judgements fall within the organically connected system of the super-ego and not within the system of the id. The id impulses, in fact, are usually in conflict with our conscience and its associated unconscious impulses, and they cannot therefore be included within any reasonable definition of the good. Again, I have at least an inkling of the suggested process of projection and re-introjection (imago formation) involved in the interaction of internal and external factors by which the super-ego is originally built up, and I am far from denying that super-egos may be cruel and tyrannical, and lead their unfortunate possessors into the madhouse. But Dr. Stephen proceeds to apply an immediate ethical judgement to super-egos at this point in the argument, without providing any firm basis on which this judgement can be based; and she is then unable at any later stage to find such a basis, but is driven to a mere assertion that the good is to be identified with the "maturity" towards which there seems to be a spontaneous tendency. My argument was that this identification of good with "maturity" can be given a rational basis if the theory of evolution is combined with that of psycho-analysis.

I begin by accepting the naturalistic view that what analysis reveals to be the content of a man's super-ego is the good as far as that man is concerned, however it may contravene other people's principles. We require, however, a wider definition of good, which applies to mankind as a whole, and which can be used to guide the efforts of the ego to assimilate its super-ego and control its id; or, in more ordinary language, can be used as a rational ideal for the personality. My contention was that just as the individual super-ego or good consists of those principles according to which the individual tries to act, so the general good consists of those principles on which the general activity of the world is based; and the main feature of the world's activity, as far as we are concerned, is the process of evolution. I am therefore stating that the super-ego is the representative within the personality of the tendencies which have been expressed in evolution. It is the business of the rational ethical thought of the ego to make the representation as faithful as possible, since I do not of course deny that the queer mechanism of imago formation does not by any means always ensure even tolerable correspondence, although natural selection will see to it that by and large super-egos do not vary too wildly. Similarly, the occurrence of deleterious genotypes is not entirely prevented by natural selection. But their existence does not hinder us from realizing that evolution depends on the properties of genotypes; and. the fact that some super-egos may be retrogressive should not tempt us to deny the evolutionary functions of that aspect of the personality.

The only alternative to this seems to be to identify that part of the personality which has produced social evolution with some system of impulses other than the super-ego. The id hardly comes into question. Nor can we plausibly attribute human evolution to the conscious workings of the ego, although we must insist that the ego should at this stage in development collaborate with the evolution-producing impulses. The final alternative is that adopted by Dr. Stephen, to attribute evolution to some otherwise unspecified "spontaneous movement". This would be inescapable if, as T. H. Huxley asserted, the character of evolution was not the kind of thing that our super-egos commonly accept as good. But this, I have argued, is not the case; and Dr. Stephen agrees in so far that she expressly defines the good in terms of her "spontaneous movement" towards "maturity". She is. thus led (a) to deny the identification of the super-ego with the good, on the grounds that some people's super-egos do not tend in the direction of evolution, (b) to invent a "spontaneous movement" in the direction of evolution, but (c) she can only point to the existence of this movement in exactly those individuals whose super-egos are not in conflict with the direction of evolution. The "spontaneous tendency" is therefore otiose; the job it is invented to do can be equally well done by the super-ego at unconscious levels, and by the super-ego aided by the ego at conscious ones. For I am of course in entire agreement with her and Professor Huxley that the conscious part of the mind (the ego) should exert itself to control, not only the physical mechanism of evolution, but, even more, the psychological mechanism whose fantastic, slapdash character has rendered man's social evolution so miserably slow and full of setbacks. But until one realizes that this mechanism does produce evolution, one is not likely to be able to assist it in doing so.

c. Letter from Dr. Karin Stephen

Actually it does seem to me that we are very near to agreement. I do not think it necessary to include diseased regressive tendencies in our idea of the goal of evolution, even if it should turn out in the end that they prevail over the forward healthy tendency. What I do wonder, however, is whether, because there appears to be a tendency in each individual personality to develop in the direction of maturity (which I have been equating with the tendency towards "good"), we can go on to argue that there is a similar tendency in the whole of evolution in that direction. I am not sure whether this may not have been slipped in unwarrantably because we use the same words, "development" or "evolution", about the personality as we do about the organic world as a whole. Is there *any* reason to suppose that evolution as a whole has the production of our "good" mature personalities as its goal? I should like to know what you feel about this. Even if there is no reason to believe this, I do not think that it upsets our criterion of good, which we can still define perfectly objectively and universally from what we observe and infer about the maturing tendency in individual human personalities. But it would make the criterion I have put forward not identical with the one you suggested and I think it would invalidate your wider evolutionary one.

About the Super-ego. I think we are sometimes at cross purposes because we take the meaning of this word in somewhat different senses. I regard the Super-ego, strictly speaking, as *always* a pathological phenomenon and therefore as *never* good, and I think this is straight orthodox Freud. The process which leads in the "good" direction ends (ideally) in the Super-ego's disappearance at the point when it gives place to regulation by the Ego. I feel pretty sure that this is not the sense in which you have taken it, and if you could put into words your meaning of "Super-ego", I think we could clear up this apparent difference—because I do not think that you would endorse your own statement about the Super-ego being the basis of ethics if you used "Super-ego" in my sense, while I should probably agree that it does apply to "Super-ego" in your sense. I am willing to agree that the Super-ego is the basis of the sense of guilt, but I do not think that guilt makes any useful contribution to goodness (oddly enough it usually seems to be an excuse for going on sinning, or a convenient excuse for cruelty). The way "badness" or "sin" or mental and emotional abnormality (synonyms on my definition of good) is overcome is not by guilt (which perpetuates them) but by a reorientation which includes an emotional change and redistribution of instinctive forces and an intellectual discovery of mistake; both, however, accomplished mainly unconsciously.

d. *Reply by C. H. Waddington*

May I take your second point first, about the meaning of "Super-ego". It is important first to consider what sort of a thing a super-ego is. You will notice that both Dr. Barnes and Professor Bernal[1] have understood it to be a distinct individual entity. And I think there is a tendency in much psycho-analytical literature to write of it almost as though it were a separate person. But I imagine that we would agree that actually it is a name for a system of mental tendencies, which hang together sufficiently to make them in some degree separable from the rest of the personality. If this is so, the limits of the system to which it is convenient to apply the name must be determined by convenience, since it is more or less arbitrary how we split

[1] See p. 114.

up the personalitv into constituent parts. Thus in my original essay I spoke of the super-ego as "a more or less isolable dynamic function within the personality"; and I wrote of ethics appearing "as the consciously formulated part of a much larger system of compulsions and prohibitions". I then went on, I admit, to use the term super-ego to cover both the unconscious compulsions, etc., and the conscious ethical beliefs which are associated with them. I suppose the orthodox psycho-analytical usage would include these conscious beliefs largely in the Ego. But what I was attempting to refer to was the whole dynamic system of the (orthodox) Super-ego plus those parts of the Ego which are associated with it. I am sure you will be able to give a more precise definition of my concept than I, as a non-psychologist, could achieve. And I think you are correct in supposing that my later development of the subject applies only to my modified concept, and not to the strictly orthodox one.

The propriety of making such a modification is, I think, made more acceptable by considerations which arise in connection with your first point, about the relation between the development of maturity in the individual and evolution in general. To my mind, the psycho-analyst's concept of the perfect personality really is by way of being an unnecessary myth. And that for two reasons. Nobody is healthy in that sense, and I have the feeling that nobody could be. Moreover, mature and well-balanced personalities are in point of fact dependant on the culture in which they live. The anthropologists Ruth Benedict, in her *Patterns of Culture*, Margaret Mead in *Sex and Temperament* and Bateson in *Naven* have· all pointed out that the type of personality who feels well-adjusted and happy is not the same in different societies. In Bali it is the warm friendly soul who becomes maladjusted and driven out of his wits, among the Arapesh the aggressive, dominating man, among the Mundugomor the unaggressive and unhistrionic. Thus a tendency towards maturity cannot, I think, be defined objectively and universally from the individual development alone, since the nature of that development depends very largely on the social order he is living in. This is, of course, essentially the same point, in a psychological instead of a political context, as I made in my reply to Bernal; observation of a single culture cannot establish a criterion by which the dominant tendency of that culture can be judged.

But, you have asked, "is there any reason to suppose that evolution as a whole has the production of our good mature personalities as its goal?" To my mind, this is putting the question back to front. It should be, "Is our culture producing personalities which develop in a way concordant with the general direction of evolution?" I rather dislike speaking of evolution having a goal. I should wish to say that we observe an evolutionary change of which we can specify certain characteristics (to take a random selection,—an increase in size, in complexity of structure, of subtlety and precision of movement, of capacity to react to the relations between objects). A tendency for change in this sense has then in point of fact been operative in the organic world. I have presented my arguments for connecting our concept of the good, through the super-ego, with the tendencies which, in human society, bring about evolutionary development. The question then is whether our society favours a type of development of the super-ego (in my sense of the word) which is congruous with the general evolutionary direction. I think one can agree straight away that evolution has in fact produced mature personalities, and that mature personalities in general are further along the evolutionary course than what you call diseased or pathological ones. The attainment of maturity of some kind would therefore be a step in the evolutionarily forward (on my definition "good") direction. But different types of mature personality may also differ in ethical value.

e. Letter from Dr. Stephen

I really begin to hope that we are now reaching the mutual understanding which would make this discussion a model controversy! I see that you object to the psycho-analysts' concept of the perfect personality as being 'an unnecessary myth', on the grounds that 'nobody is healthy in that sense and . . . nobody could be'. I agree that you are quite right in saying that such perfection is impossible, but I do not see that that is any argument against defining the notion. It seems to me rather like objecting to setting up such absolutes as 'black' and 'white' in a world composed of shades of greys, or to defining a straight line in a space where nothing is perfectly straight, though lines can approximate more or less to straightness. The definition, though hypothetical, is useful. If we think of the development of human personalities as showing movement in either of two directions, forward towards maturity, or backwards in the direction of repression and disintegration, then I see no objection to the 'perfect personality' as the hypothetical goal of this forward movement. Your objection to speaking of evolution as having a 'goal' follows, I expect, from the same line of argument. The imaginary goal of evolution is a myth in the same sense, but I suggest, a hypothetical assumption is useful for clarifying our ideas about the direction in which evolution appears to be going. Like so many other scientific hypotheses it is founded on all the particular observations we have succeeded in collecting of a movement or tendency in a certain direction of change. It is only objectionable if we fall into the mistake of imagining that it is a real observed fact itself.

I believe that the rest of our apparent disagreement depends on words, that is on the meaning we propose to give to the notion of the Super-Ego. When I said that my definition of it was straight orthodox Freud I think this was not completely accurate because, actually, no single defini-tion can apply to all the ways in which Freud uses the term. What is meant by the Super-Ego depends on what level of development is being considered. In all cases, how-ever, I agree with you in regarding it as a dynamic function within the personality. My definition applied to the very primitive levels of Super-Ego functioning in infancy and early childhood, when it starts as little more than a protective reflex recoil from shock, suffering and terror at being overwhelmed, expressed in denial and flight. In its somewhat later stages Freud thinks of it more in terms of introjection of or identification with the parents (or their substitutes) who are used by the child, as it were, to supplement its own inadequate power of regulating itself. At this stage these introjected powerful Beings are thought of as directing the child's behaviour from inside according to standards which the child imagines the parents demand from it (which is often a very different thing, in practice, from what they do really expect, because the child's sense of reality is still highly fantastic). Rather later still the Super-Ego is thought of as the internal need to conform to an external standard set up by the parents, or parent substitutes, which the child may feel it ought to follow from various motives. It may *have* to submit, against its will, from fear, or it may *want* to be what they want it to be from love and admiration. In the case of *having* to conform through fear the Super-Ego is in conflict with the rest of the child's self and is generally a harsh tyrant against whom it rebels secretly. Its energy is borrowed from the child's own aggressive, destructive, revengeful impulses, directed back against itself to deprive it of satisfaction. This 'bad' Super-Ego is the motive force behind the Puritanical Conscience for which all pleasure is sinful. It usually goes along with a 'moralistic' attitude towards the rest of the world also, whose effect is to grudge other people their happiness and to try to punish them.

In the other case, where the motives for conforming to the standards of the Super-Ego are love and admiration, the energy is borrowed from the child's love-impulses and the standard is more kindly and tolerant and permissive, both towards itself and towards other people.

The particular code of behaviour enjoined varies with the standards set up by the parents (or other admired models) and with the prevailing culture; it may be exacting and may demand renunciations, but the child's wish to obey is whole-hearted and not divided against itself by secret rebellion. It is a love-Super-Ego, and it might even be described as 'free', 'voluntary' self-regulation following a model which the child has assimilated and made its own, in contrast with 'compulsive' submission to an external law which is imposed by the other kind of hate-Super-Ego. Willing obedience to a love-Super-Ego is sometimes described by Freud as following the Ego-Ideal, and it might be useful to adopt this word and keep the word Super-Ego for the punitive kind of conscience which is motivated by fear. The pursuit of the Ego-Ideal would then be 'good' because it would produce the sort of harmonious, integrated personality which seems to be the goal of personal evolution in the direction of maturity, whereas the Super-Ego, at all its levels of development, would be 'bad' because it drives the personality to be in conflict with itself and so disintegrates it.

If we may divide the internal force which says 'I must' into these two different sorts of conscience, or moral urges, we can now, perhaps, reach final agreement. You, I think, would agree with me in not regarding the fear-hate-Super-Ego which is obeyed only under compulsion and disrupts the personality, as the basis of all moral action and in calling it 'bad': I would agree with you in regarding the pursuit of the Ego-Ideal, which is acceptable to the whole self and integrates it, as 'good': and, moreover, these meanings of 'bad' and 'good' will give us the objective criterion of Ethics for which we are looking which will be valid in all circumstances, no matter what the divergence of codes of 'right' and 'wrong' behaviour may be in different cultures.

I should like, in conclusion, to attempt to complete this effort to establish peace and good will by trying to show that this definition of the meaning of 'good' not only reconciles Dr. Waddington and myself, but even goes a considerable way towards reconciling us also with several others of the contributors to our discussion, though I am not blind to the fact that the agreement must be only partial.

I even begin to hope that Professor Bernal and I, for instance, might be able to shout to each other a few words of fraternal greeting across the immense gulf which divides us. Professor Bernal's objection (p. 114) to our bringing in the Super-Ego, on the grounds that we are attempting to explain the better known by the less known, has already been answered by Dr. Waddington, but I should like to question, modestly, whether Professor Bernal has not under-estimated our knowledge about this despised 'myth', or hypothesis, as I should prefer to call it. We know really as much about it as Newton knew about gravity when he observed the falling apple, and, actually, these two hypo-thetical forces, gravity and the Super-Ego, have about equal claims to our attention, since both enable us to formulate under general laws the behaviour of events which, without them, would be inexplicable and appear chaotic.

But our possibilities of agreement go further than this. I am able to agree with Professor Bernal's attack on external sanctions for ethics so long as he directs them against the idea of sanctions from On High, but I believe I can detect, even in his own writings, an underlying appeal to some absolute standard which is not claimed as Divine but which he believes to be inherent in the laws of economic 'progress', and which, I suspect, he assumed as self-evident and there-fore has omitted to formulate explicitly. When he says that 'the sanctions of ethics are imposed through the conditioning process to which every human being is subjected, etc.', he is talking, of course, only about the code of behaviour held to be 'right' or 'wrong' in individual cultures and, as I have explained, I have no idea of claiming that any of these are reliable guides to the meaning of good since they can, and do, contradict one another. Professor Bernal points out that old values are replaced by new ones, but what he means by saying that 'the recognition of this gives us the unifying principle which Waddington's analysis still lacks', I am at a loss to understand, unless he means that none of them are wrong but each is right at different stages of social evolution. I cannot believe that this is his meaning because it would involve him in maintaining that 'the essential commandments of the capitalist epoch . . . respect for property, sexual morality conditional on property relations and the avoidance of thoughts likely to upset the social structure' are right so long as Capitalism is still in being, and I do not believe he would maintain this. Moreover, in any case, as I say, I seem to be able to read between the lines an implicit assumption of ethical values whose validity does not get its sanction from any economic situation which may happen to prevail at any particular period, but which, on the contrary, make some economic arrangements 'better' (ethically) than others.

These values, he tells us, were prevalent 'in the earliest Societies' which 'have given rise to the most deeply and commonly felt virtues . . . compassion, comradeship, fair-dealing'. He tells us further that we are now being driven (by dialectical materialism, I expect) to return to these virtues in the modern form of 'a strong sense of human equality' and 'social responsibility', to be followed by 'other virtues we cannot see now'. What these virtues will be of course he is not in a position to say, at present, but I feel pretty sure he envisages them as conducing to greater co-operation and greater harmony.

I doubt whether anthropologists would agree with this rosy account of the Golden Age of virtuous primitive Society, and some students of psychology might question whether the economic forces which are disrupting Capitalism are in fact to be relied on to secure for humanity the virtues which Professor Bernal foreshadows, but, at any rate, there seems to be little doubt that he considers compassion, comradeship, etc., ethically superior (and absolutely so) to the qualities produced under Capitalism and this is what interests us at the moment. Indeed, he admits himself that 'by introducing the ideas of efficiency and harmony' he is 'in fact still invoking ethical judgments'. And why should he not? I hope he will be as delighted as I am to discover that his ethics are no less absolute than ours, and that, in fact, they boil down to precisely the same standards as my own, the only difference being that I describe the achievement of this 'good' in terms of the growth of the human personality towards maturity through the replacement of Super-Ego compulsion by Ego self-regulation, while he, I suppose, believes that it will be brought about by the agency of dialectical materialism. This being only a matter of our personal views concerning means, perhaps Professor Bernal will shake hands with me over the final goal which we shall both welcome as absolutely 'good' (or at any rate absolutely 'better' than previous achievements and tending in the direction of 'goodness'), though before I will grasp the hand I must ask him to eat his words and admit that he too does believe in a tendency towards some goal (which, admittedly, does not exist now and which may be incapable of achievement and certainly is at present unknown), but which can only be described in terms of 'good' personalities and which he (and I) would both agree in calling absolutely 'good'. As regards what the 'goodness' of the personalities would consist in he is not very precise and we can only guess from hints dropped about harmony and comradeship and efficiency. I should hardly have thought efficiency could be regarded as 'good' in itself, but only as a means which can be used to produce 'good' or 'evil'. I am able to be rather more explicit in my account of what 'good' maturity produces in the personality because I have frequent opportunities of observing this very process of growth at close range in patients who make progress in it in the course of treatment. The essential points are that it brings about internal harmony and self-regulation, thus increasing internal security and so making it possible to outgrow the fantastic picture of omnipotent, dangerous agencies, inside and outside the self, which belongs to the magical state of immaturity, and to replace it by a more realistic view. This increase in the sense of reality reduces the tendency to react with fear and rage to others and with Super-Ego frustration to the self, with the result that the personality becomes more courageous and able to make safe contact with life and to develop the judgment needed to direct its behaviour intelligently so as to secure real satisfaction. This success sweetens the character, reduces hate, envy and greed, and liberates love and generosity. These human characteristics are what I mean by 'goodness' and I shall be surprised if they do not coincide with those tacitly assumed by Professor Bernal to be inevitably forced upon us by economic necessity and that is why (though I disagree with his idea of how we are to come by them), I claim that we are, fundamentally, in agreement.

And now, finally, I am in hopes of grasping another hand, that of the Bishop of Birmingham. If I understand his parable (p. 127) rightly he means that Professor Joad and I are only dressing up in a language of modern metaphors the age-old controversy about whether or not the Devil exists, and that I, all unknowingly, find myself on the side of the devils (who, of course, imply the angels too). That we are still discussing the same phenomena of human suffering which, under the name of sin, was debated for many centuries by the Fathers of the Church I warmly agree, but I believe we are now making a little progress, after a long and somewhat weary detour. I agree that we have got back to recognizing that the struggle underlying these strange and painful phenomena, which I also look upon as moral problems, is between non-material entities, but whereas the Fathers believed that the participants were separate Beings located outside the victim, we now place them within and regard them as parts of the patient himself. Nevertheless I do believe that the Fathers of the Church and I have some real common ground in our views on ethics. That the modern explanation which I have tried to put forward may still be no more than metaphor I will concede, if my opponents will equally recognize the metaphorical nature of their own explanation. I do not think we need be ashamed to do this because scientific hypotheses are in fact all based on metaphor. What else is the wave theory of light, or the old atomic theory which described the structure of matter in terms of the behaviour of minute billiard balls? Metaphors are selected to throw light on the laws governing the changes which take place in the less known by comparison with the better known and they vary with different stages of our knowledge. The old anthropomorphic God-Devil metaphor was the obvious one to use when the Church began to wrestle with the problem of what we now call intra-psychic conflict and it is interesting to find that, even to-day, many of the insane (I say this without disrespect) revive this old explanation of evil influences and possession to make some sense for themselves out of the horrible experience of being governed by unknown forces which control their thoughts and actions against their own conscious wills. The psychiatrist no longer believes the patient's explanation and holds that the unhappy man is suffering from a splitting of his own personality in which one set of impulses, or constellation of hopes, fears, and beliefs, has got cut off from the rest and is no longer in his conscious awareness or subject to his voluntary control. What we may be saying about it all 1,000 years hence there is no telling, but that the agency of unconscious impulses is a useful hypothesis which explains a lot of things that do happen is undeniable. Take, for instance, the case of a woman who was hypnotized and told that the hair brush lying on the dressing-table had become so heavy that she could not possibly lift it, and was then woken up and asked to pick up the brush. To her amazement she found she could not, try as she might. The doctor could see the reason. Although, with one hand, she was pulling as hard as she could, with the other, all unknowingly, she was pressing the brush firmly on to the table. The struggle ended by her wrenching the handle off the brush! She believed it was bewitched. The doctor preferred to say that she acted out of unconscious obedience to his hypnotic suggestion. This is a trivial illustration not involving 'good' and 'evil', but I agree that, in some of the conflicts which may wrack the human personality, the struggle between the two is joined. The Bishop of Birmingham and I have at least that much in common: we also agree over the importance of resolving the conflicts and establishing the triumph of the 'good', however we may differ as to the means to be employed in bringing this about. And I have small doubt that we agree also as to the characteristics which I, in the name of maturity, and he, under a different name, hold to be 'good'.

Perhaps it may be suggested that this fundamental agreement about what human attributes are 'good' which exists, as I have dared to hope, between people so different as Dr. Waddington, Professor Bernal, the Bishop of Birmingham and myself, should make us suspicious of the whole thing, since we have all been brought up to believe that these things are 'good'. If we had belonged to a culture which glorified and valued envy, malice and all uncharitableness, if such a culture there be, might we not have discovered arguments in support of their 'goodness' instead?

If what the anthropologists tell us is true it is very disturbing, but is it true? Perhaps the savages have been pulling our legs: perhaps our translations of their words do not convey their meaning . . . there have been those who took their theistic beliefs as proof that all human beings have an innate awareness of God, but when they say God it may be questioned whether they mean what the Bishop of Birmingham means, and when they say a man is 'good' may they not quite possibly mean only that he is powerful and terrible? It is even possible that those savages whose culture, we are told, so contradicts our notions of 'good' may have grown up under conditions so adverse that they have never got near enough to maturity to know 'goodness' at all.

Until the anthropologists have disposed of these doubts perhaps we need not allow ourselves to be too much intimidated by them.

f. Reply from Dr. C. H. Waddington

I only wish to add a very short note to Dr. Stephen's last letter. As she has pointed out, our views are very similar. But I still think that she does not appreciate the full force of the anthropological evidence that the conception of "good" differs in different societies. The anthropologists have not merely asked various tribes what they consider to be good; they have observed different societies to determine what demands of conscience are in fact generally approved of. It is, I think, impossible to suppose that the differences they describe between cultures depend merely on difficulties of verbal translation. They are real facts, which have repercussions for instance in the psychological realm in which Dr. Stephen is mainly interested. The type of personality which is driven, by its maladjustment to society, over the border line of madness is not the same amongst the histrionic and treacherous Iatmul described by Bateson in *Naven*, the moody and suspicious Dobuans studied by Fortune, the paternal and co-operative Arapesh of Margaret Mead or the stolid and ritualistic Zuni of Ruth Benedict.[1] The Super-Ego can only become "mature", or, as Dr. Stephen phrases it, can only become an Ego-Ideal, in the form which society approves. Our society approves (perhaps only rather formally) an Ego-Ideal which can be called a Love Super-Ego. Other societies do not, and a Love Super-Ego attempting to develop in them is deformed by social pressure into an unbalanced, even lunatic, personality. I am, at least in a general way, in agreement with Dr. Stephen, that the Love Super-Ego approved (rather half-heartedly) by our society is good. In this opinion I may be deceived by the fact that I have grown up within a society which to some extent approves that ideal; but I am urging that the basis on which my opinion should be formed, and from which I claim to be able to derive it, is the congruity of that ideal with the general course of evolution.

[1] See G. Bateson, *Naven*; Margaret Mead, *Sex and Temperament*; R. Fortune, *Sorceress of Dobu*; R. Benedict, *Patterns of Culture*.

2. A COMMENT BY MRS. MELANIE KLEIN

Dr. Karin Stephen has stated lucidly some aspects of the psycho-analytic position. There are, however, sides of this problem which she did not cover, and which seem to me pertinent both to the understanding of the origin of the super-ego and to Dr. Waddington's thesis.

Here in brief outline are some of the facts which have become clear to me in my psycho-analytic work with young children, and which I wish to bring to your notice. The feeling of 'good', in the baby's mind, first arises from the experience of *pleasurable* sensations, or, at least, freedom from painful internal and external stimuli. (Food is therefore particularly good, producing, as it does, gratification and relief from discomfort.) Evil is that which causes the baby *pain* and tension, and fails to satisfy his needs and desires. Since the differentiation between 'me' and 'not-me' hardly exists at the beginning, goodness within and goodness without, badness within and badness without, are almost identical to the child. Soon, however, the conception (though this abstract word does not fit these largely unconscious and highly emotional processes) of 'good' and 'evil' extends to the actual people around him. The parents also become embued with goodness and badness according to the child's feelings about them, and then are retaken into the ego, and, within the mind, their influence determines the individual conception of good and evil. This movement to and fro between projection and introjection is a continuous process, by which, in the first years of childhood, relationships with actual people are established and the various aspects of the super-ego are at the same time built up within the mind.

The child's mental capacity to establish people, in the first place his parents, within his own mind, as if they were part of himself, is determined by two facts : on the one hand, stimuli from without and from within, being at first almost undifferentiated, become interchangeable ; and on the other, the baby's greed, his wish to take in external good, enhances the process of introjection in such a way that certain experiences of the external world become almost simultaneously part of his inner world.

The baby's inherent feelings of love as well as of hatred are in the first place focussed on his mother. Love develops in response to her love and care; hatred and aggression are stimulated by frustrations and discomfort. At the same time she becomes the object upon whom he projects his *own* emotions. By attributing to his parents his own sadistic tendencies he develops the cruel aspect of his super-ego (as Dr. Stephen has already pointed out); but he also projects on to the people around him his feelings of love, and by these means develops the image of kind and helpful parents. From the first day of life, these processes are influenced by the actual attitudes of the people who look after him, and experiences of the actual outer world and inner experiences constantly interact. In endowing his parents with his feelings of love and thus building up the later ego-ideal, the child is driven by imperative physical and mental needs; he would perish without his mother's food and care, and his whole mental wellbeing and development depend on his establishing securely in his mind the existence of kind and protective figures.

The various aspects of the super-ego derive from the way in which, throughout successive stages of development, the child conceives of his parents. Another powerful element in the formation of the super-ego is the child's own feelings of revulsion against his own aggressive tendencies—a revulsion which he experiences unconsciously as early as in the first few months of life. How are we to explain this early turning of one part of the mind against the other—this inherent tendency to self-condemnation, which is the root of conscience? One imperative motive can be found in the unconscious fear of the child, in whose mind desires and feelings are omnipotent, that should his violent impulses prevail, they would bring about the destruction both of his parents and of himself, since the parents in his mind have become an integral part of his self (super-ego).

The child's overwhelming fear of losing the people he loves and most needs initiates in his mind not only the impulse to restrain his aggresson but also a drive to preserve the very objects whom he attacks in phantasy, to put them right and to make amends for the injuries he may have inflicted on them. This drive to make reparation adds impetus and direction to the creative impulse and to all constructive activities. Something is now added to the early conception of good and evil: 'Good' becomes the preserving, repairing or re-creating of those objects which are endangered by his hatred or have been injured by it. 'Evil' becomes his own dangerous hatred.

Constructive and creative activities, social and co-operative feelings, are then felt to be morally good, and they are therefore the most important means of keeping at bay or overcoming the sense of guilt. When the various aspects of the super-ego have become unified (which is the case with mature and well-balanced people), the feeling of guilt has not been put out of action, but has become, together with the means of counteracting it, integrated in the personality. If guilt is too strong and cannot be dealt with adequately, it may lead to actions which create more guilt still (as in the criminal) and become the cause for abnormal development of all kinds.

When the imperatives: "Thou shalt not kill" (primarily the loved object), and "Thou shalt save from destruction" (again the loved objects, and in the first place from the infant's own aggression)—when these laws have taken root in the mind, an ethical pattern is set up which is universal and the rudiment of all ethical systems, notwithstanding the fact that it is capable of manifold variations and distortions, and even of complete reversal. The originally loved object may be replaced by anything in the wide field of human interests: an abstract principle, or even a single problem, can come to stand for it, and this interest may seem to be remote from ethical feelings. (A collector, an inventor or a scientist might even feel capable of committing murder in order to further his purpose.) Yet this particular problem or interest represents in his unconscious mind the original loved person, and must therefore be saved or re-created; anything which stands in the way of his objective is then evil to him.

An instance of distortion, or rather reversal, of the primary pattern which at once presents itself to the mind is the Nazi attitude. Here the aggressor and aggression have become loved and admired objects, and the attacked objects have turned into evil and must therefore be exterminated. The explanation of such a reversal can be found in the early unconscious relation towards the first persons attacked or injured in phantasy. The object then turns into a potential persecutor, because retaliation by the same means by which it had been harmed is feared. The injured person is, however, also identical with the loved person, who should be protected and restored. Excessive early fears tend to increase the conception of the injured object as an enemy, and if this is the outcome, hatred will prevail in its struggle against love; moreover, the remaining love may be distributed in the particular ways which lead to the depravation of the super-ego.

There is one more step in the evolution of good and evil in the individual mind which should be mentioned. Maturity and mental health are 'good', as Dr. Stephen pointed out. (Harmonious maturity, however, though a great 'good' in itself, is by no means the only condition for the feeling of adult 'goodness', for there are various kinds and orders of goodness, even among people whose balance is at times badly disturbed.) Harmony and mental balance—furthermore happiness and contentment—imply that the super-ego has been integrated by the ego; which in turn means that the conflicts between super-ego and ego have greatly diminished, and that we are at peace with the super-ego. This amounts to our having achieved harmony with the people whom we first loved and hated, and from whom the super-ego derives. We have travelled a long way from our early conflicts and emotions, and the objects of our interest and our goals have changed many times, becoming more and more elaborated and transformed in the process. However far we feel removed from our original dependencies, however much satisfaction we derive from the fulfilment of our adult ethical demands, in the depths of our minds our first longings to preserve and save our loved parents, and to reconcile ourselves with them, persist. There are many ways of gaining ethical satisfaction; but whether this be through social and co-operative feelings and pursuits, or even through interests which are further removed from the external world—whenever we have the feeling of moral goodness, in our unconscious minds this primary longing for reconciliation with the original objects of our love and hatred is fulfilled.

3. *a*. A COMMENT BY MISS MIRIAM ROTHSCHILD

Although the difficulty of being brief without inferring too much and explaining too little is very great, I nevertheless would like to mention two small points in connection with this unusually interesting discussion.

Firstly, I query Dr. Waddington's and Dr. Stephen's suggestion that one might characterize the "good" tendencies of super-egos as "healthy". On the contrary I believe that large numbers of persons, who have unquestionably contributed to the forward movement of human evolution and thought, have manifested diseased super-egos. Thus, to mention the first which come to mind, J. J. Rousseau was an obsessional neurotic, at times actually exhibiting "the curious phenomenon of compulsive behaviour", Van Gogh a schizophrenic, Swammerdam a melancholic, Ampère an unusual type of psychopath, John Clare a maniac depressive. True, some of these individuals landed themselves in lunatic asylums, from where they painted or wrote some of their greatest works. They certainly spread infinite unhappiness among their immediate human contacts, but few would deny their contribution to human progress.

In no scientific field can Professor Haldane's final remark apply so forcibly as in the field of psychology. Not only is this relatively new study shrouded in an obscure jargon, but also in ignorance. In the present state of our knowledge, it appears a little doubtful if we are capable of evaluating the extraordinary super-ego. To many of their contemporaries the great ethical thinkers (or "mutants") themselves appeared as diseased super-egos, representing as Dr. Stephen says, a "menace to humanity" and were dealt with accordingly. The asylum authorities used Van Gogh's paintings to fill up a hole in a wall—an act which appeared to them as anything but antisocial.

Secondly, students of animal behaviour—particularly of social animals—are well aware that various "moral codes" are inherited in a very cut and dried fashion. Thus a fear of trespassing is inherited in the Black-headed Gull as part of its territorial instinct.[1] A trespassing bird behaves in a most peculiar way, displaying in an exaggerated manner those characteristics which in human behaviour might be termed the visible and outward signs of a thoroughly guilty conscience. I feel that from a Black-headed Gull's point of view Professor Joad must be unquestionably right and there can never be a moment's doubt about the dictates of their conscience vis-à-vis the rights and wrongs of trespassing. As Dr. Stephen expresses it: the still small voice is a raging dictator. This type of territorial behaviour, modified to suit the social life of these birds, no doubt has survival value in the avoidance of conflict. One can easily conceive, that should the Black-headed Gull evolve an even closer type of social structure their Seventh Commandment would change along the lines indicated by Professor Haldane from "Thou shalt not trespass" into "Thou shalt not stake out territories".

[1] Lecture, Institute for the Study of Animal Behaviour, December 1938.

I feel that the contributors to this discussion have under-estimated the hereditary element in our ethical codes, and have unduly emphasized the role of individual psychological types, experience, reason, etc., etc. Some of the most striking phenomena in animal behaviour are those inherited *trends* of behaviour which require a relatively very small amount of conditioning in order to fix them.

It is fairly certain that it will take relatively much longer for the Black-headed Gull to change its Seventh Commandment than for man. But presumably when this takes place the bird's whole organism will be adapted to their new outlook. Genes carrying the desire for territories will have been eliminated. We are less fortunate, and even if our minds have reached the stage when we know we should live as Christ enjoined we are thwarted by relics from a not distant and very extensive past, when such an awkward charac-teristic as for example the impulse to dominate was finally and securely fixed by natural selection. It was pointed out in a poetical way in Genesis that the unique evolutionary feat of man, namely the development of a self-conscious and reasoning ego, was not conducive to a smooth passage through time. Can science cope with the fact that, mentally, man has raced ahead of his genes? In the answer to that question I believe one may ultimately find the true relation-ships between science and ethics.

3. *b*. REPLY BY DR. C. H. WADDINGTON

Miss Rothschild's first point, in which she draws attention to the contributions made to man's ethical development by people who were themselves unbalanced, is, I think, a considerable difficulty to any theory which attempts to base ethical judgements on considerations relating to isolated individuals. Dr. Stephen's theory that a harmonious super-ego is synonymous with good could perhaps be expanded to cope with the difficulty by introducing distinctions between still more aspects of that highly complex entity, a human personality; but the expansion would not be altogether easy. Miss Rothschild's difficulty is, however, quite easy to surmount on my theory which relates the good to the general course of evolution. For it is clear that the course of evolution is primarily an affair of whole species or societies. It is, in fact, just because of his contributions to the progress of mankind in general that we value such a man as Van Gogh, and recognize as less important the fact that he was a plague to most of his after all not very numerous friends. It is exactly in approving, rather than disapproving, such characters, that the evolutionary criterion differs from the more individualistic criteria on which most ethical systems are based. From the evolutionary point of view, we should regard it as being probably generally advantageous that a society should contain many different types of person, and even an occasional madman. But, of course, it would be nonsense to invert these considerations and attempt to establish the current ethical values of the mass of mankind by reference to the super-egos which have governed the behaviour of the few madmen of genius; a society of Van Goghs, Beethovens and Rousseaus would be inviable from the start.

It is of course true that the social value of these mad geniuses is very rarely recognized during their lifetime. This is unfortunate; but it is not a theoretical difficulty in the way of accepting my formulation. If it is a difficulty for any theory, it is for those systems of thought which were actually employed in condemning them; and the theory of the super-ego has at least the negative excuse that it had not been invented at the relevant periods of history. But, more than that, it does, as we have just seen, make it possible to classify them as valuable, whereas it is very doubtful whether the more orthodox theories can find any convincing way of saving their face even *post hoc*.

Miss Rothschild's second point, as to the importance of the hereditary determination of our ethical standards, is, to my mind, very much less convincing. We actually know very little about the causal mechanisms of behaviour in birds, and the mere invocation of a genetic basis whose expression cannot be modified by the environment seems somewhat too easy. What, for instance, is supposed to have happened to the herring gulls which have recently taken to living inland away from sea coasts? A mass mutation? What of the differences in behaviour of sea birds in isolation and among the crowds on the nesting sites? What of the cormorants of Holland, which, in that rockless land, nest in trees, unlike their fellows in any other part of the globe? The reference to Black-headed Gulls is in fact not only a long range and dubious analogy, but is an analogy with something which we understand even less fully than the human behaviour which is supposed to be illuminated by it. I hope Professor Joad will not misunderstand me if I say that the statement that a Black-headed Gull would agree with him seems to me an unjustifiable insult to the bird.

But if we dismiss the birds to their own rookeries, the possibility remains that man's ethical behaviour may be determined by his genes. Bateson, in the early crusading days of genetics, had no doubt about it: "It is upon mutational novelties, definite favourable variations, that all progress in civilization . . . must depend".[1] "So soon as it becomes common knowledge—not a philosophical speculation but a certainty—that liability to a disease, or the power of resisting its attack, addiction to a particular vice, or superstition, is due to the presence or absence of a specific ingredient; and finally that those characteristics are transmitted to the offspring according to definite, predictable rules, then man's views of his own nature, his conception of justice, in short his whole outlook on the world, must be profoundly changed."[2]

Now in my view a mutational novelty may have ethical value; it may make possible the development of an individual who is better than any of the unmutated type. Such

<hr />

[1] William Bateson, *Naturalist*, p. 353. [2] *Ibid.*, p. 328.

a possibility certainly arises if one accepts the theory advanced here that the general direction of evolutionary advance defines the meaning of "good"; whether it can arise on any alternative theory which makes the direction of evolution irrelevant to the nature of goodness, I am not prepared to state. But on my theory one is bound to admit that the genetic nature of mankind sets limits to the range of human ethical performance, just as does that of chimpanzee to his. But that does not in the slightest imply that differences within the range of normal human behaviour are also genetically determined. That would only follow if we supposed that the environmental differences met with among men have only negligible effects on the development of the different genotypes. Bateson seems to have made such a supposition; and so, I think, does Miss Rothschild, although she supplements it with the remarkable, and I am afraid impossible, suggestion, that "mentally, man has raced ahead of his genes". But the evidence is all against the ineffectiveness of the environment; there is no space to review it here and I may perhaps refer the reader to my recent textbook[1] where fuller references will be found.

The environment *may* produce differences of the same order of magnitude as those which we are used to finding among normal men and which we are concerned with in a discussion of ethics. Equally, one cannot deny that genetic differences may be responsible for variation within the same range. The determination of the parts played by these two agencies in the actual variation among men is a difficult matter. The theory which attributes the major importance to heredity is to-day known as the Race Theory—and it numbers some fairly disreputable customers among its friends. Again the evidence is too involved even to summarize here. One can only say that the facts make it impossible to deny the important part played by the environment, whereas there seems to be no compelling reason to attribute much weight to the contribution of heredity; and one can

[1] *Introduction to Modern Genetics.*

refer to Muller's famous essay on "The Dominance of Economics over Eugenics". The conclusion one must come to is that we cannot shelve the problems of ethics by referring them to the uncontrollable mutations of our genes.

CHAPTER 5

SOME PHILOSOPHICAL DIFFICULTIES

a. Letter from C. H. Waddington to Professor H. Dingle

DEAR PROFESSOR DINGLE,—

I have just returned to Cambridge to find the proof of your comment on my article in *Nature* which was presumably sent to me instead of to you by an error on the printer's part.

May I give the gist of what I should like to reply? My main thesis was, as I thought, the opposite of apriorist; namely it was that ethics are derived from experience. Your point, as I understand it, is that my assertion is not derived from experience and has no consequences in experience. To take the first part of this first. I believe I might reply by the simple *tu quoque*. Your assertion that my statement is clearly not derived from observation is itself not derived from observation. But if this rather school-boyish reply does not satisfy you, I think I should proceed as follows. The derivation of a philosophical assertion such as mine from experience is not easy, at any rate to me, to understand, but I still believe that such statements have a meaning. It seems to me that they probably cannot be derived from any contemplation of experience, but only from participation in it. By participation in it I mean essentially attempting to alter it. And thus the first part of your criticism, as to derivation from experience, dissolves into the second, as to the application to experience. In that connection I think my reply is clear. If my thesis is correct, the considerations which should be taken into account, in determining whether it is ethically right to bomb Germany, are those relating the presumed results of that action to the general progress of human society; the people best competent to discover the facts and assess them are military scientists, sociologists and historians. If the contrary thesis about ethics is correct, the considerations to be taken into account are intuitions, and those most competent to consider them are priests, poets and prophets.

b. From Professor H..Dingle

I don't want to bother you unduly with my views, but I should like to say that I dissent not merely from the particular ethical principle that you suggest but, more fundamentally, from the general idea that ethics can be treated as a science. It seems to me to belong to an essentially different class of problems. Science can tell you what to do to reach a desired end; it cannot tell you what end to desire, and that is the root of the problem. You say "the considerations which should be taken into account in determining whether to bomb Germany is ethically right are those relating to the presumed results of that action to the general progress of human society". But what is the "progress" of human society? Opinions differ. Still more fundamentally, is the progress of human society desirable, or the production of a Nietzschean superman for which society should sacrifice itself? Almost anyone can answer this question, but no one has yet given me any reason for his answer that is not merely a dogmatic statement that it must be so. Not very scientific, that. (Incidentally, your evolution criterion might be used very effectively to back up Nietzsche.)

A still more general difficulty is contained in the phrase "presumed results". The characteristic of most actions having an ethical quality is that they have to be performed in ignorance of the results. If I see a drowning man, I have no time to enquire whether he is Hitler or Christ: I have to act at once and my act is good or bad irrespective of its consequences. Such a situation never arises in science because what makes an action scientific is not the action itself but what you conclude from it. If you omit to read the galvanometer at the right time the experiment may be ruined, but you can repeat it. If it is an astronomical event that doesn't recur, you are still not unscientific unless you draw false conclusions; you are simply an inefficient observer, and science is not violated but only retarded. But whatever ethical principles one holds must be independent of the results of one's actions because they are nearly always unknown.

I should like to take up your reply about the relation of your assertion to experience, but have already wearied you enough.

c. From C. H. Waddington

You said that the root of the matter is that science cannot tell you what end to desire. I think that view derives from considering a desire as an *a priori* subjective thing with no roots in experience. I should say that in ethics we discuss "myself acting to attain an end", and that an analysis of this into "myself with a desire" and "an act aimed at that desire" is invalid. It is invalid because the desire has been formed by the impress of experience. Consider the following analogy. One is turned loose in a factory devoted to making musical instruments; after a time one discovers that the thing to do in such a place is to put the bits together so that they make violins, trumpets, etc. And this would not depend, as I see it, on having *a priori* interest in music. Similarly ethical ends are determined by what the world is actually like. The realization of this is concealed by two main factors; firstly, by the fact that man is concerned with a system having two orders of complexity, of individual and society, just as the man in the instrument factory might get confused between considerations of making single instruments and considerations of making orchestras. And secondly, by the peculiar emotional tone of compulsion which characterizes early-acquired ethical knowledge, the reasons for which I discussed in my original paper. But, strong though the compellingness of ethical commands may be, it is in fact very seldom a categorical imperative. We may be in comparative ignorance of the results of our acts in ethically-difficult situations, but I don't think we are indifferent to them. We rescue drowning men because any arbitrary individual is more likely to be a useful member of society than the reverse. Further, I suppose that many of our ethical beliefs have been perpetuated by natural selective forces (working with a good deal of imprecision, I admit) and ethical principles may have been built up on a basis of their results independently of whether anyone can in fact predict these results with any accuracy. It may be that societies in which drowning men are not automatically rescued induce mental attitudes which lead to disaster when the society is submitted to particular stresses. The ethical belief in rescuing people would then be produced, whether anyone understood the reason for it or not.

d. From H. Dingle

And now to the ethical question. I expressed myself badly in speaking of a desire and an act aimed at it; I should have spoken of "purpose" and an act aimed at it; no thought of emotion was in my mind. I do not know if you would maintain that a conscious purpose was determined by experience. I would say that it was not, though I would agree that, *in retrospect*, all one's behaviour might be correlated so as to appear determined. That, however, raises the wider question of free-will, but it bears on an objection I have to your analogy; namely, that examination of the instrument factory can only tell you what the parts were presumably made for; it does not limit your freedom to put them together to form better (or, if you like, simply other) articles. Ethics is essentially directed towards the future, and I see no reason why it should be limited by the past.

Let me put the matter in another way, starting from scratch. We must act (passivity counts as an action since it influences things generally) and we usually have a wide choice of actions. This, I take it, is common ground. Next, the choice we make is not a matter of indifference; hence we need some guidance in making it, and that guidance we call an ethical principle. (The logical positivists, I think, would have to deny this, though I have not observed that they act purely at random.) The problem is then to determine ethical principles, and I see no unexceptionable way of solving it. You say the course of evolution dictates them. That to me is simply a dogmatic statement. Consider the following objection to it. It is only within the last one hundred years that we have known of the existence of the course of evolution, let alone what it .has been. What was "good" for people before that time? They needed guidance as much as we. It seems to me that you must say that they had merely to guess, and that people who were then what we now call "moral" were either so by accident (this takes some believing) or else were guided by something valid other than the course of evolution. If you choose the latter alternative, I should ask why that guidance isn't available now, and why it should not take precedence over the course of evolution, particularly as we may always make discoveries showing that the course of evolution is other than we now believe.

e. From C. H. Waddington

To return to your argument about ethics. I do mean to assert that an aim or purpose is, in the last analysis, determined by experience, although of course in particular cases the experience may have been that of many generations back, the purpose being transmitted by the process of teaching.

If I may follow your argument-from-scratch, I think I can indicate where and how I diverge. I follow you in saying that we must act, that we have a range of choices open to us (or at least appear to have) and that our choice is not a matter of indifference. As you see, this has already raised the problem of free-will, which I don't claim to be able to solve; you will have seen my remarks on the subject in my reply printed in Chapter 2 (p. 40). For present purposes I shall accept the above statement at its face value.

In discussing how we make our ethical choice, you seem to me to suggest that a man wakes up in front of a problem with a completely empty mind, that he has to start looking for a set of ethical principles, and, as you admit, you do not see what he can base them on. I should say on the other hand that a man in such a situation *has* a set of ethical principles; the question is where did he get them from. My answer, or rather the answer of the psychologists, is that he learnt them at a very early age from his parents, etc. For any individual man they may be anything you please. But for man in general they cannot be independent of experience; they, like all the rest of his nature, have been selected to enable him to survive. Natural selection is of course a statistical affair, and many individuals and groups of individuals fail to pass its test. If therefore man, now being conscious, wishes to apply his intelligence to bettering his chances of progress, he can discover which principles of action have in fact been successful so far by investigating the course of evolution. There is, as far as I can see, nothing at all dogmatic in this. It only appears dogmatic if one thinks of ethics as something which one already knows about, and which I am, for obscure reasons, asserting to be identical with the principles of evolution. But, instead of that, I claim that ethics are acquired principles according to which one tries to change the world; and I am pointing out that the principles by which the world has been changed are exemplified in the course of evolution.

The validity of ethical systems, according to the above discussion, may be independent of their conscious intellectual derivation. It is assured by natural selection. Ethical progress without any sound intellectual understanding is no more impossible than was evolution in general without understanding. But once the possibility of understanding either ethics or evolution arises, I do not see how one can suggest that the old "intuition" (i.e. unformulated analysis) should take precedence over the conscious investigation.

f. From H. Dingle

We seem to be dealing with two different problems. You ask, "how we make our ethical choice", by which you mean "how have we made our ethical choice". That is a matter of rationalizing past experience; it is a task for psychology. But my problem is different. It is not how to explain what has been done, but how to decide what to do now.

The first problem is an academic one—very interesting, but not, in my view, properly called an ethical problem. I agree with you that it is a branch of science; I disagree that it is ethics. We may discover why Charlotte Corday killed Marat in terms of family training, etc., but we have not thereby discovered whether her action was good or whether she deserves praise or blame for it. You may say this is a meaningless question. If so you would agree with the logical positivists, and I am not sure you would not be right. Not being a psychologist, I am not very concerned about it, one way or the other.

But the second problem is a vital one. I have to choose my actions now. You agree that the choice matters, and I cannot evade responsibility for it on the ground that I am not a psychologist or anthropologist. How am I to choose? That is the problem of ethics, and it is essentially different from the problem of some future investigator who, when my action is a *fait accompli*, may try to determine why I chose it.

I therefore disagree with your statement that "a man in such a situation has a set of ethical principles, the question is where did he get them from". *If* he has such a set, then that is *a* question, but the *ethical* question is: "Are the principles which (supposing I have them) I got from my parents or elsewhere the principles which I should allow to determine my actions?" It is only because people have faced that question that ethical progress has been possible.

I cannot see how a designation of past principles as "good" is going to lead to their supersession.

g. From C. H. Waddington

You suggest there are two different problems involved in our discussion; 'how did I make my ethical choice" and "how shall I make it now". I cannot see that there is any real distinction here. Your letter suggests to me three possible ways in which one might try to draw such a distinction, but I do not believe that any of them can be sustained.

1. One might suggest that when investigating the past one can determine the reasons why a certain person adopted certain methods to attain their ends whereas the important problem of the present is to discover what ends to strive for. But it seems to me clear that theoretically one can investigate the reasons for past choices of aims just as much as of means.

2. The difference between past and present might be made in terms of free-will; that my past actions seem to have been determined whereas my present ones, I feel, are freely chosen. Again, my view does not, I think, involve me in asserting this; according to it, past choices and present choices have the same degree of freedom.

3. It might seem that I suppose that past choices are taken with reference to persistent ideals which were adopted in early childhood and have remained unmodified, whereas present choices are taken, partly at any rate, in accordance with a process of reasoning. This again is not involved in my view. I quite recognize that reason is involved in the formation of ideals or aims, at all stages later than the very earliest.

All these possible grounds for drawing a distinction between past and present choices seem to me, in fact, to be possible grounds for *feeling* that there is a distinction, but not logical grounds for making a *rational* distinction. It seems to me quite impossible to maintain that the processes by which I shall make a choice now are different in kind from those by which I determined an ethical aim in the past. In both cases my aim will be formed in response to my experience; according to observation, my earliest experience (family life, etc.) will be of profound importance, but will be modified, through the agency of reason, in the light of later experience. The task of a theory of ethics is to clarify and systematize the data of this later experience and to draw the correct deductions from them. When I suggest that (a) the function of that part of the personality to which we owe our idea of the good is to enable human societies to exist and also to evolve, and that (b) the direction of that evolution can be roughly specified; these statements are both scientific, in that they are based on observation of past events, and pertinent to a present attempt to modify our ethical goals in the light of reason. The fact that the same statements can be relevant to both your problems shows, I think, that there is no real distinction between them. Your example of Charlotte Corday's action in killing Marat was not a fair statement of the case because I should of course agree that one cannot base any general ethical standards on a single act; my statements above only acquire their relevance to my present ethical choice because they are generalizations which refer to ethical choices in general.

h. From H. Dingle

I think I should agree with you in rejecting your three proposed distinctions between the problems of rationalizing past and present ethical choices (with a possible reservation with respect to (2)). A more valid one, I think, lies in the fact that we can consider past choices in the light of what has followed as well as preceded them, and detect unconscious influences, whereas we cannot decide present actions on other than conscious grounds. But in point of fact, that is not the *kind* of distinction I am making. My distinction is not between the means available for rationalizing past and present actions, but between the rationalizing of any action and the determination of what action to make. In the first problem the action (or choice) is a datum, just like the observational phenomena of any other science. In the second it does not yet exist, and the whole problem is to make it. This distinction seems to me fundamental, and I cannot see how the second problem can be solved, or even approached, by the scientific processes appropriate to the first.

I have long felt that this ethical question has got to be threshed out much more carefully than has hitherto been done, and I am very glad to hear that you may be writing a book on it. Your attitude seems to me to be the one most likely (if any attitude is) to lead to something definite being arrived at ultimately. I may add that I have no theory of my own up my sleeve, and am not at all concerned to refute your views through any belief in revelation or anything else as the proper source of moral judgments. I simply do not know how to decide what it is "right" to do in given circumstances, and although it is a psychological fact that I prefer Christian ethics to Nazi ethics, for example, I cannot prove that I am right in acting in accordance with that preference.

i. From C. H. Waddington

I of course agree with you that there is a valid distinction to be made between "the rationalizing of any action and the determination of what action to make". But it seems to me these two do not differ through and through in the nature of all the processes involved; the difference is rather that when making a present choice an act of will is added on to rationalization-process of the same kind as is involved in analysing a past action. It may be true that in analysing the past one can, because one has more leisure, more easily discover all the factors, both material and of an unconscious nature, which have played a part in the decision to act in that way. But in making a present choice, one at least attempts to take all such factors into account. The situation seems to be exactly similar in other non-ethical contexts.

One can analyse fully some physical phenomenon of the past, whereas one cannot be sure, before one has made the actual attempt, whether a proposed course of physical action in the present has been based upon an adequate analysis. One might say that one can only have a physical science of the past, not of the present; and in a sense that would be true. But it would not enable one to deny that physical science, although only capable of strict application to past events, is the best guide, indeed is developed solely in order to be a guide, to the courses of action to be chosen in the present. Similarly, I should not wish to suggest that one can hope to find an infallible ethical rule; my suggestion is primarily concerned to indicate which factors are most relevant to an ethical choice, and secondarily to point out that certain generalizations about their action may be possible.

j. From H. Dingle

My reply to your last comment would be that I do not agree that the difference between (a) the rationalizing of a past action, and (b) the proper determination of a future one is expressed by (b) = (a) + an act of will. The analogy with ordinary scientific procedure seems to me a false one. Knowledge previously obtained is indeed the best guide to what experiments to perform now, but that is an incidental aspect of it. If we perform an unsuitable experiment, then (unless, of course, it happens by pure accident to lead to a new discovery) we simply ignore the result and perform another one. Science is not in the least benefited or injured by our waste of time, because its essence is in what we learn from experiments, not the particular means by which we learn it. But the ethical character of an act belongs to the act itself and not to its results. An act, acknowledged by the doer and others to be bad, may have good results, and *vice versa*. Indeed, it seems to me part of our proper business to see that the consequences of evil acts are good. I would not, of course, contend that the consequences of an act, so far as they can be foreseen, are entirely irrelevant to its ethical value; that would be absurd. But what I say is that they are not of its nature, and that an act is as fully good or bad when they cannot at all be foreseen (as when a man fights for a cause whose triumph is completely uncertain, and the conditions are such that if he succeeds things will be much better, and if he fails, much worse, than if he didn't fight) as it is when they can.

In short, your remark, "I should not wish to suggest that one can hope to find an infallible ethical rule", virtually concedes my point, for that is just what one must find—or, if not a rule, at least a means of knowing definitely that what one does in each particular act is the right thing for him then to do. That does not mean a means of knowing what the consequences of his act will be. It is desirable to know that, and a scientific approach will be necessary, or at least desirable there. But the ethical character of the act depends on other things. It is a theoretical problem to construct an ethical system for omniscient beings. The practical problem is to find one applicable to a particular person as he is, here and now.

I have formulated my objections to your argument in the following paragraphs, which I intended for publication in the correspondence columns of *Nature*. I learn, however, that the Editor now desires to close the controversy.

Dr. Waddington now makes it clear that he uses the term "ethical problem" in an unusual (and, I think, inaccurate) sense. An ethical problem I take to be the problem of deciding, in a given situation, which of the possible actions one should choose. Dr. Waddington takes it to be the problem of explaining why one has, in fact, chosen a particular action. The latter I call a psychological problem, and when he says that my view of ethical principles "discounts at the outset the possibility of observing the genesis of aims", he is incorrect. I admit that possibility but, like some others, I wonder what it has to do with ethics. Ethical judgments, which are Dr. Waddington's data, are my goal.

Dr. Waddington objects to my statement that "the principles of action must in essence be independent of the consequences of action", but I think I can confirm it by an example not too remote from the recent experience of many Europeans. A system which a man regards as good is attacked from without. If he successfully resists the attack the consequences will be better than if he assists it or does nothing. If he unsuccessfully resists it, the consequences will be worse than if he assists it or does nothing. He estimates his chances of successful resistance as fifty-fifty. What is he to do?

I would observe that (1) the alternatives mentioned are exhaustive—he *must* choose one of them; (2) the choice must be made in ignorance of consequences. I freely admit that, after he has chosen, his mental activity might be describable in terms of egos, super-egos and the rest, and perhaps evolution might not be irrelevant; and for all this it is quite unimportant which choice he makes. In the meantime, however, the ethical problem stands: what is he to do? The choice there is not unimportant.

k. From C. H. Waddington

I am very interested in your new formulation of your point of view, but I personally find it at least as unusual as you seem to find mine. Recent thought about ethics has so completely accepted the relevance of the anthropological data about the different views various civilizations have as to the nature of the good; and been influenced so profoundly by the psychological investigations into the processes by which we do in fact form our ethical judgements, that I think your dismissal of all this material as beside the point will strike many besides myself as very strange and difficult to grasp. Thus your rejection of the idea that "observation of the genesis of aims" has anything to do with ethics presumably implies that men's actual aims are never ethical, but merely sometimes (by chance?) approximate to the real ethical aim, whose validity is independent of them. Such a view asserts that when we are confronted with a choice of action, the ethical qualities of the various choices are simply characteristic of the actions, fixed for them by God or by the structure of the universe, but quite independent of the man who is making the choice. To my mind, discussion of ethics in those terms is either, as you sometimes partly admit, quite impossible, or at least purely metaphysical. But in either case it has nothing to do with ethics in the practical sense, which is concerned with how we do in fact determine what we consider to be an ethically good aim. The impossibility of discussing the matter in such a formulation arises, I think, because the formulation is itself nonsense; it is surely clear that we do meaningfully discuss whether an action is good or bad, and any formulation which makes it impossible to do that is thereby shown to be incorrect.

I also oppose your example given in the last two paragraphs of your letter. Even if we assume, as you do, that the consequences of losing this war will be much worse than never having fought it, and that our chances of winning are fifty-fifty, surely it is clear that the grounds for deciding to fight it are still estimates of the future consequences. It cannot be in itself good to murder a large number of Germans; it can only be so if it is estimated that there is a reasonable chance of bringing about a very much better result. The decision of whether it is right to take the risk depends on a calculation, however imperfect, of what is at stake and what the odds are. If one sees someone on the point of being swept over Niagara Falls, it is not merely silly, but also, in most cases, wrong, to commit suicide by jumping in "to rescue them".

As man is a social being, one must always bear in mind the importance of inspiring examples, and it is probable that it is usually right to accept short odds. You state that your example is going to demonstrate the irrelevance of consequences to ethical value. But what you actually do is merely to assert that the ethical choice about this war is unimportant for the personalities of the people involved and for evolution, both of which statements seem to me totally untrue; and you then assert that there is some other unspecified way in which the choice is important, but as you don't specify what this important relation is, you are left with the rhetorical and unanswerable question of how important the choices are.

l. From H. Dingle

I should not have thought my view unusual, whether right or wrong. Surely most people would agree that an ethical problem is a problem of deciding how to act (usually in difficult circumstances), and that this is quite a different problem from that of analysing past actions, though the solution of the latter, of course, may be an important factor in solving the former. Nor is it unusual to hold that the rightness or wrongness of an act is independent of its consequences (though I did not say that, but only that, since we rarely know the consequences, our decision how to act must be taken on independent grounds). The triteness of Tennyson's

> Because right is right, to follow right
> Were wisdom in the scorn of consequence

and the Christian admonition to walk by faith and not by sight, is surely evidence enough that it is not unusual to regard the practical guide to action as independent of consequences.

However, it is not of much importance whether either of us holds unusual views: the important thing is whether they are right or useful. To me the problem is how to act from day to day, and I regard it as a severely practical problem, which has to be solved at each moment with the knowledge I possess. It may be that if I knew everything about the genesis of aims and the consequences of each alternative, I should have solved the problem. The fact is, however, that I don't, and I can't wait until I do before deciding how to act. Hence I must have some *working* principles of action, and it is those that are the essence of the problem.

That is the essential difference between this and an ordinary scientific problem. In the latter, one can suspend judgment when there are insufficient facts, or make a working hypothesis solely for the purpose of getting more facts and discard it when its work is done. In ordinary affairs, however, judgment can't be suspended, and one cannot, for example, kill all crooners in order to test the plausible working hypothesis that the world would be more pleasant without them. The act of applying the hypothesis has itself an ethical aspect here, whereas the act of testing a scientific hypothesis has no scientific aspect apart from the conclusions drawn from it.

m. From C. H. Waddington

I have just read, not only your last letter, but the whole of our correspondence. I think the arguments which cause you to reject the possibility of a scientific theory of ethics emerge quite clearly from it. May I bring our interchange of letters to a close by stating, as clearly as I can, the reasons why I feel unable to accept your arguments?

As I see it, you have advanced two main theses. In your last letter, you point out that "the essential difference between this (i.e. an ethical problem) and an ordinary scientific problem" is that in the latter "one can suspend judgement when there are insufficient facts", while "in ordinary affairs judgement can't be suspended". That is of course quite true of science in the laboratory. But judgement can also not be suspended in respect to non-ethical problems arising in ordinary life. Suppose that, for some reason, one had to make a bridge to get to the other side of a river, and that the only materials available were ones whose strength one did not know. One would have to act as best one could with what knowledge was available; but surely one would not doubt that the essential theory which one required was a scientific one. I should in fact claim that this is always true of all applications of science, however well-accepted the theory may be; and I should deny absolutely that the residual uncertainty robbed science of its fundamental importance as the essential guide to practical behaviour. It is an applicability of exactly the same kind as this that I wish to claim for a scientific theory of ethics.

I do not wish to minimize the fact that situations in which one is forced to act on a quite obviously inadequate basis of fact are much commoner in ethical situations than in many other spheres of activity; nor, as I admitted to the Dean of St. Paul's, do I wish to deny the value of many of the intuitively formulated ethical teachings of the past. But I do not agree that the imperfections of scientifically based ethical theory are sufficient ground for rejecting the possibility of such a theory.

I cannot even agree that ethics would have to be classed, with geology and astronomy, among the non-experimental sciences. Such drastic experimental procedure as the slaughter of all crooners is not, I admit, either practicable or desirable. But very many experiments with different ethical hypotheses have in fact been made by different societies or social groups; they range from large numbers of individual trials of aberrant views by "social misfits", through group attempts such as many progressive schools, to large-scale experiments such as the different political systems. I should suggest that some previously held ethical beliefs, such as "an eye for an eye", or the acceptance of slavery, have been rejected on the experimentally determined grounds that they do not work out in practice.

Your second thesis seems to be most clearly stated in your letter No. *j*, in which you state that what we must find is "an infallible ethical rule . . . or if not a rule, at least a means of knowing definitely that what one does in each particular act is the right thing for him then to do". The argument is presumably that as we often do not know enough to reach anything approaching certainty as to the results of our actions, we must discover some infallible *a priori* rule. But clearly this is not, logically, at all necessary; we can content ourselves with forming the best working hypothesis we can. And I suggest that the latter is what we in fact most often do. It is only this necessity to form working hypotheses on inadequate data which leads to the great psychological conflict which is such a prominent feature of many important ethical choices.

I suspect that this argument of yours was not put forward simply on the flimsy logical grounds I have just mentioned, but is, partly at least, based on the observation that some ethical beliefs do seem to have absolute validity. This I have previously admitted, and I have suggested grounds of a psychological nature from which this appearance of absoluteness may be derived. In doing so I am definitely implying that an appearance of absolute validity is not a safe guide to the choice of ethical principles. Your statement that what we must find, when confronted with an ethical choice, is an infallible rule, seems to me to be calling for an impossibility; since it is only because not all our ethical beliefs appear absolutely valid that ethical choices arise. In fact, I think one could probably even go further, and state that a powerful feeling of infallibility only attaches to beliefs which are actually the subject of psychological conflict; is there not perhaps the paradoxical situation that we are only tempted to insist on the universal validity of a belief when we are, in some not very conscious way, not quite sure whether it is correct? Be that as it may, I cannot accept a search for universally valid rules as necessitated either by logic or by inference from the observed characteristics of those ethical beliefs which we tend most generally to accept. In my opinion an ethical choice can only be settled by enquiring into the probable consequences of the various courses of action and into the derivation of the ethical motives by which one is tempted to guide one's behaviour.

n. *From H. Dingle*

I do not think that the two "main theses" you attribute to me properly represent my view. I do not wish to withdraw from them, but the first is merely incidental, introduced at a late stage in a particular context, and not a "main thesis". The second is more fundamental, but I think you have misunderstood it since you refer to "universally valid rules". I did not ask for such rules, but only for a means by which the individual can know what "in each particular act is the right thing for *him* then to do".

This has been a long correspondence, and since we are no nearer agreement on the fundamental issues I think it might be well if I give the essence of my view in what I would regard as two main theses, so that you can the better locate your divergence. Our common ground (including agreement that "our choice is not a matter of indifference") is shown in my letter *d* and yours *e*.

(1) The ethical quality of an act is concerned entirely with the motive of the act, and not with its physical nature. The consequences of the act, however, depend on its physical nature; hence they are irrelevant to the ethical quality. I grant that my estimate of the consequences will determine the physical nature of my act, and that I can, and should, make that estimate on scientific lines.

(2) A motive implies an object aimed at. Hence the fundamental ethical question is "Towards what general object should my acts be directed?" The answer "Towards the continuation of the course which evolution has taken in the past" evokes the question "Why? On what grounds should I adopt this rather than 'Towards the perfection of my individual self'; 'Towards the supersession of the cosmic by the moral law'; 'Towards the realization of the aspirations of Jesus Christ or Gautama Buddha or any of the existing Churches'; etc., etc.?" It is that question to which I cannot see the possibility of a scientific answer.

o. From C. H. Waddington

I should like to express, as shortly as you have done, the points on which my opinions differ from those formulated in your last two points.

1. I cannot accept a fundamental distinction between motives and the physical consequences of acts, since I hold that motives are ultimately determined by the nature of the physical world. The consequences of an act thus belong to that very category of things from which motives are derived, and they are therefore entirely relevant to the making of distinctions between motives, that is to ethics.

2. The question "Towards what general object should my acts be directed?" has no meaning unless one can elucidate the significance of the word "should", which is impossible until the question has been answered; it is in fact a rhetorical expression of the question "Towards what general object *are* my acts directed"? And my reply is that, *in general*, the objects towards which my acts are directed are those which can be deduced from an observation of the course of evolution. The ethical aim is no more a matter of free choice, uninfluenced by material considerations, than is the respiratory choice (oxygen, though some parasites get along after losing their respiratory apparatus) ; or the sexual choice (reproduction, though dandelions use their sexual organs purely for show, and reproduce by other means).

CHAPTER 6

A MARXIST CRITIQUE

THE UNITY OF ETHICS

BY PROFESSOR J. D. BERNAL, F.R.S.

THE discussion which Dr. Waddington has started turns upon two questions which he asks at the outset: "Why do we feel in ourselves that anything is good?" and "Is there a Good outside and independent of what we feel?"

These questions he answers by equating the feeling of goodness to satisfaction of the demands of the Super-ego, and defining objective Good as the Direction of Evolution.

Now, while this analysis represents a considerable advance on the views of the past and present theologians and moral scientists, its advance from that position is only negative. It sweeps out of the way certain venerable ideas, but nevertheless, it still represents the same form of thinking that it criticizes, and reproduces in a new language the essence of many of the old ideas.

The essentially logical criticism of the formulations put forward is that they attempt to explain the familiar by the unknown. The feeling of goodness is referred to the Super-ego, objective goodness to the direction of Evolution. Now this is certainly better than explaining them as the voice of conscience and the will of God, because it does not introduce the essentially anthropomorphical view of the Universe, which we now know cannot have any objective validity. Nevertheless, the nature of the Super-ego and of the direction of Evolution, although legitimate objects of study, are not in fact known. In the first place. as we can see from Dr. Stephen's contribution, the nature of the Super-ego is still subject to very considerable controversy, even among psychoanalysts. It is not at all clear that the Super-ego does provide a sufficient source for ethical feelings but whether it does or not, the relevant point here is that it has been introduced into psychology largely in order to account for the existence and nature of these feelings. To use the Super-ego as Waddington does to explain those feelings, is to add to entities without necessity and, by Occam's razor, nothing is to be gained by invoking it. The same argument applies to the absolute definition of Good, given in Waddington's words as "the direction of Evolution is good because it *is* Good". Now, however much we may know about evolution, it is quite clear that we can know less about its direction than we can about the immediate past and present state of human society. To appeal to this Direction as a standard to apply in the present may be emotionally useful but cannot be intellectually maintained. Indeed, what it seems to me that Waddington has done— in line with many reformers of ethics of the past—is that he has thrown away old myths and sanctions, but has felt the necessity to introduce new ones for justification.

The danger of introducing myths is that they tend to crystallize lines of thought and, therefore, prevent development. The whole history of modern science has been that of the struggle between ideas derived from observation and practice, and pre-conceptions derived from religious training. It was not, as we so often think, that Science had to fight an external enemy, the Church; it was that the Church itself—its dogmas, its whole way of conceiving the Universe —was within the scientists themselves. The distortion of viewpoint was not externally imposed, it grew up internally in the mind of each thinker.

A characteristic of this distortion was its duality—the distinction between the world of fact and a world of values. This duality is most explicit with Descartes, and it should be noted that though both the material and spiritual worlds were originally compatible with the anthropomorphic

Universe, the material world was always several steps ahead in its liberation from this view. After Newton, God ruled the visible world by means of Immutable Laws of Nature, set in action by one creative impulse, but He ruled the moral world by means of absolute intimations of moral sanctions, implanted in each individual soul, reinforced and illuminated by Revelation and the Church. The role of God in the material world has been reduced stage by stage with the advance of Science, so much so that He only survives in the vaguest mathematical form in the minds of older physicists and biologists. In physics He is needed only to explain the creation of a Universe, which is discovered, as research advances, to be less and less like the one with which we are familiar. In Biology He is invoked to account both for the origin of Life, and for the general Purpose of Evolution.

Now the history of scientific advances has shown us clearly that an appeal to Divine Purpose, or any supernatural agency to explain any phenomenon, is in fact only a concealed confession of ignorance, and a bar to genuine research. Accordingly, whatever he may be off duty, the average scientist is not a conscious theist in his work, but he may be, very often, an unconscious one.

Originally, the Universe was supposed to exist as an expression of Divine Will and Purpose, or Providence. Those who still think so are less muddle-headed than those who talk of purpose without admitting the necessary implication that purpose can only have meaning to a personal God. The process of deanthropomorphism discussed by Professor Fleure has still a long way to go. The division between the world of fact and the world of values is still thought to be fairly sharp. The relation of Science to Ethics is seen as a definite and limited one. Scientific knowledge is of use to find the means for achieving good things, but it has nothing to do with the determination of what is Good. In this view, many of the scientists as well as of the theologians who have contributed to the present discussion are in full agreement. The great merit of Waddington's contribution is that it goes far beyond this untenable half-way stage. He sees that Science must have something to say not only on how values can be achieved, but also on our appreciation of values, and on the validity of the values themselves. Nevertheless, by erecting the separate myths of the Super-ego and the Direction of Evolution, he still retains something of the original dualism.

The way to a new and unified conception of ethics will only be open if we are prepared to abandon altogether the requirements and justifications that apply only to the older views. Waddington feels the need for an absolute justification of ethics and finds it in the Direction of Evolution. Now this is essentially an act of Faith. We cannot, as I have already pointed out, *know* enough of the direction of Evolution to base any useful deduction as to present ethical standards, but we can, choosing our ethical standards from our experience of present-day society, and our knowledge of its development, project them into a direction of evolution, and then invoke the Direction of Evolution to justify these standards. Stated in this way, the process seems palpably nonsensical, if not dishonest, but its emotional value is also manifest. What we are doing in a more refined way is exactly what the primitive tribe did in creating its tribal god and then rigorously enforcing his commands on the members of the tribe. If we reject such justifications we must, it seems to me, logically also reject the demand for any external or cosmic standard of ethics. We will find in making this rejection that we have lost nothing but our illusions.

External sanctions for ethics are intellectually empty. If we remove them we see the real sanctions inside Society itself. For the last few hundred years the character of these sanctions has been determined by the class nature of society. The sanctions have been essentially conservative. The essential commandments of the capitalist epoch out of which we are just now emerging were: respect for property, a sexual morality conditional on property relations, and the avoidance of thought likely to upset the social structure. Marx and Engels, as long ago as the Communist Manifesto, pointed this out; in the dissolution of present-day society, it is now becoming apparent to everyone.

The sanctions of ethics are imposed through the conditioning process to which every human being is subjected from the moment of birth, in family upbringing, in education, in initiations into social life. The conscious sanctions of the morality and law are secondary ratifications to repress deviations from a norm that is internally and unconsciously acquired. This complex of individually acquired ethics retains relics of all the successive systems of ethics that have been fitted to the conditions of earlier societies and economic systems. The earliest societies have given rise to the most deeply and commonly felt virtues—compassion, comradeship, fair dealing. At various later periods and for different groups, current ethics has contained elements more or less at variance with these and with each other. These contradications have mirrored in the human mind the external conflicts of society. The tasks of the great ethical reformers have been to formulate attitudes and feelings that would remove the internal conflicts but, until Marx, few had seen that this could never succeed without tackling their external causes.

There is no fear that criticizing away the theological or philosophical bases of ethics will lead to the total loss of ethical values. In fact, the close interdependence of modern human groups has vastly increased the necessity for ethics. The destruction of one set of values only occurs as new ones take their place. We can observe this happening in vastly different directions in Nazi Germany and the Soviet Union. The social basis for morality cannot be abrogated. It may, indeed it must, change with the changed conditions of society, but its continuous existence is bound up with that of society itself. The recognition of this is the unifying principle which Waddington's analysis still lacks. It underlies implicitly most of his arguments, but is prevented from full expression by the abstraction of the Super-ego and the Direction of Evolution. To a certain extent this is conceded in his first reply to his critics. In this he amplifies his original definition of good as the direction of evolution by pointing out that "if the ethical system is to be derived from the nature of the external world, we must pay attention to what that world is like; and one of the most important data is the scientifically ascertained course of evolution". Similarly, the Super-ego becomes simply a term for the sum of the social impress on the individual affecting his behaviour and his feelings towards that behaviour. These are far less metaphysical claims than the original one and open the way to an appreciation of the relation of science and ethics.

The part that Science has to play in ethics is not in the determination of absolute values or, with the theologians, as a mere means of achieving eternal values laid down from on High, but rather the interpretation and understanding of a changing society, so that the inherent socially induced ethical motives in mankind can work themselves out most effectively and harmoniously.

Now by introducing here the ideas of efficacy and harmony I am, in fact, still invoking ethical judgments but these, at least, can be expressed without myths, in terms of social and individual behaviour. The first contribution of Science to social development was that which provided increased control over environment. The economic requirements underlying the origin and development of human society—namely collaboration to secure more abundant and reliable food supply and general conditions of well being—can in modern industrialized communities be fully secured only by application and development of natural science. But human social development has notoriously never permitted the full utilization of technical powers. The basic reason for this has been social conflict. Society from the very start differentiated itself locally into tribes, nations and states, and socially into classes. The individual and collective pursuit of well-being has led to an endless struggle between such groups and between individuals in them throughout all history. Human technical efficacy is persistently frustrated by the lack of social harmony.

True, in the past, it was largely as a result of such clashes that technical progress took place, but at a price that it can no longer be borne. We have now reached a critical turning point in social development. The scope and power of social organizations is so great that conflict is already reversing the trend towards better material conditions and threatens the very lives of the majority of mankind. It is only by understanding the nature of the social and economic development that we can prevent this happening, and this, we are now learning, is also the task of science, particularly of the social sciences. Up to now, the official social sciences have lamentably failed in this task because they themselves have been subjugated to the very forces which have produced the major disharmonies in social development. Effective understanding has come far more through the combination of theory and practice, exemplified in the Marxist development of the Soviet Union.

It is here at last that we touch the fundamental change in the conception of ethics. Ethics is not something that can be abstracted and set apart from practical human life. If we attempt to do so we can only perpetuate tragic delusions of the past. We do not need any such abstract conception as the Good. What we need is a full understanding of the world in which we live, intimately coupled with our own personal and social activity in changing that world. The Marxist view that right action and right understanding are inseparable has always been a stumbling-block to philosophers and moralists of the old school, unable to think of human behaviour except in terms of anthropomorphic abstractions such as Will, Conscience or Desire, but it is far more in harmony with modern anthropomorphical and psychological views. Once it is grasped the problems of ethics, morality, politics and economics are seen as one general social problem which is being dealt with in the measure that it is understood.

Ethics will remain personal in the sense that the task of each will be different and limited by their understanding and the influences which have worked on them. It will be objective in the sense that human organizations—trade unions, research institutes, armies, the whole people—are consciously striving together for immediate needs that are seen clearly and following out a directioned development which is realized more and more clearly as it is vigorously pursued.

We are living through the major crisis of a great period of transition in human history. A new society is growing and struggling in the matrix of the old. When it becomes dominant it will be found that a new ethics has grown up along with it. New human values will be found taking the place of some of the older values and reinforcing others. A strong sense of human equality, social and economic as well as political, will be a primary virtue, the practice and consequently the true understanding of which is impossible in our class and race-ridden world. Social responsibility and co-operativeness will take the place of prudence, abstinence, and exclusive concern for family and dependants. The scientific attitude and increased consciousness of the structure and development of society will take the place of piety and respect for tradition. There will be other virtues, the nature of which we cannot see now because they have no field of application in present society. But we need no knowledge of these developments to know what we value and strive for in the present. Absolute values are as illusory as *a priori* knowledge. Ethics which is an expression of human society has as infinite a future.

2. REPLY BY DR. C. H. WADDINGTON

Professor Bernal's very important comment on my essay brings into clear focus two of the points which I had con-

sidered of most importance. He suggests that I have introduced two new myths, of the super-ego and of the Direction
of Evolution; concepts which he, unlike myself, dignifies
with capitals.

As to the first of these there is, I think, no real disagreement between us. Bernal begins by claiming that I have
introduced the super-ego as a new *ad hoc* entity invoked to
account for our feelings of the good; and with a flourish
of Occam's lethal snickersnee he despatches it. But then he
finds himself left with—what? Why with just that "social
impress on the individual" which I had originally indicated
by that term. Exactly what mythical intruder he thought
he had glimpsed in my first essay—exactly whose head rolled
so ignominiously in the dust—remains to me at least a
mystery.

But although my use of the term super-ego did not, I
think, introduce any new and supererogatory concept, it did
involve something more than the employment of new name
for an old idea. Bernal suggests that we know much less
about the super-ego than about our feelings of the good.
In a sense that is true—but only in the sense in which we
always know more about the immediate data of perception
that about the scientific concepts into which we analyse
them. This process, far from providing a basis for an "essential
logical criticism", is the foundation of the whole scientific
method. Its justification is that although we may at first
sight appear to "know" more about apples and chairs than
about vitamins, essential oils and the cohesive forces between
molecules, we know it less relevantly. Similarly, I should not
like to claim that we know much about the super-ego; but
I should argue that we know extremely little about the
causal antecedents of our ideas of the good. And by classi-
fying these antecedents among the entities investigated under
the name of the super-ego, I was bringing them into relation
with a body of data with which they are not always connected. As Haldane pointed out, Marxist theory has been
concerned more with the history of ethics in the develop-

ment of societies than in that of individuals, and it needs to be amplified by some account of the relation of ethical systems to the other aspects of personality. Again, classical Freudian psycho-analysis, owing to its concentration on pathological conditions, and its neglect of the external social forces impinging on the individual, has not formed an entirely satisfactory picture of the structure of the ethical beliefs of normal people; I have discussed in more detail with Dr. Stephen the propriety of enlarging the meaning of the term super-ego to cover these more normal mental functionings. In both respects I was suggesting amplifications of the current theories; but not amplification in the sense of introducing a new entity.

Bernal's second criticism is more fundamental. He suggests that the real sanctions for ethical behaviour are inside society itself, that we cannot know enough about the direction of evolution to use that as an effective standard, but that we do in fact derive our ideas of the good from observation of society and then project them into our knowledge of evolution. Now again, all these points have a considerable measure of truth in them, but they fail to meet the point completely. I agree that our ideas of the good are, *in the first place*, imposed through "the conditioning process to which every human being is subjected"; I agree that the "conscious sanctions of morality and the law" *may be* "secondary ratifications to repress deviations from a norm that is internally and unconsciously acquired". I agree that people *may*, and in fact have, as Engels pointed out, transferred moral ideas derived from society on to their exposition of evolution. But I do not agree that these things must be so. Throughout at any rate the later part of life, the conscious and rational functions of the mind attempt to modify the early-formed and partly unconscious functional systems so as to adjust them more adequately to a wider body of data than were available at the time of their formation. This, formulated by the psycho-analysts as an incorporation of the super-ego and id into the ego, is something quite different to the mere expression in overtly rational terms of the promptings of the internal mental forces. There is, or there can be, a real transcending of the early impressions which are based on a naïve reaction to social forces; and it is with this attempted adjustment that an intellectual philosophical discussion of ethics is concerned.

Granted that such intellectualization is possible, it might still be true that, as Bernal argues, it is from contemporary social history that our most objective and unassailable ethical data might be drawn. It is to be noted that Bernal does not wish to omit all reference to historical development, as, perhaps, Spinoza might have done. His basis is recent social history of mankind, mine is the whole history of living matter; the difference is one of time-scale. Greater objectivity must, I think, be attributed to the larger time scale simply because it includes the smaller. It is always possible that the last few hundred years of development have been retrograde; as they would be if, for instance, a world fascist state came into being. Bernal's criterion would then give him a definition of good which, I should claim, was objectively incorrect when tested by my criterion. Moreover, it is, as Bernal himself so eloquently points out, exactly in order to determine the ethical status of various social alternatives that we at present most pressingly require an objective criterion; and it is probably theoretically unsound, and certainly very unconvincing in practice, to suggest that the criterion can be found within the very system to which we wish to apply it. In point of fact, people do not derive the same ideas of good from their observation of recent history. In theorizing from such observations they are meeting their infantile super-ego on its home ground, where it nearly always wins. Bernal could probably convince me with his case for deriving the values of "efficacy and harmony" from the development of capitalist Europe; but a Visigoth of the time of Rome's decay, an Aztec of Chicken Itza, a Hindu in the time of the Moghuls, might have deduced quite other values from their own civilizations, and as far as I can see

Bernal would have no grounds on which to combat their ethical ideas.

Finally, it may be admitted that an observation of the whole course of evolution does not issue in such detailed conclusions as would the study of present-day society. But it is necessary to remember that we are investigating evolution only in order to determine a direction, not to receive detailed instructions. Once we have decided from evolutionary data whether the continuation of man's progress demands a high degree of respect for the individual rather than his thorough subjection to a group-organization for instance, we can work out the implications of that directive in present political terms. Only when we have found our Ten Commandments in general evolution, can we discover our Deuteronomy in political analysis.

CHAPTER 7

SOME FINAL COMMENTS

1. CONCLUDING REMARKS BY THE BISHOP OF BIRMINGHAM

As I have read the very interesting discussion to which Dr. Waddington's essay has given rise, I have reflected that two tendencies, both dangerous, seem to beset men of science, and particularly psychologists, in their consideration of ethical and philosophical problems. They use metaphors too frequently and they create entities too readily.

Take Julian Huxley's phrase 'the repressive mechanisms of the unconscious'. Here 'the unconscious' is an entity. Mach would therefore remind us that it is a compendious mental symbol for a group of sensations. What are these sensations? and are we quite sure that they warrant us giving independent status to the entity? Further, in applying such a phrase as repressive mechanisms to an aspect of mental activity, a metaphor is used which passes from mind to matter with a facility which somewhat disquiets me. The disquiet becomes bewilderment when I read that energies can be locked up, or released from their grapplings, and that vital human impulses can be dead-locked by misguided repressive super-egos. Even 'blind decontrol' is metaphorical, though perhaps not quite as metaphorical as 'blind automatic repression'. When Waddington, commenting on Engels, writes of 'unrestrained biological drives' in connection with 'the imperativeness of the socially determined Good' he does not, to my thinking, bring light to our darkness.

It may, of course, be urged that no harm is done by such lavish use of metaphor and that we know full well what we mean when we create such an entity as the 'super-ego'.

I have doubts and, if Dr. Stephen will accept a good-natured jest, I would point my doubts by a spurious fragment of history:

Early in the second century of our era Joadus, a distinguished Stoic philosopher, criticized adversely the belief in devil-possession then held by Christians in common with the majority of men. He asked: "What . . . does all this talk about the devils and their imposition upon the personality . . . really amount to?" In reply, Karina Stephena, a cultured Alexandrian lady of Christian sympathies, said that if Joadus would study the curious pheno-menon of compulsive behaviour, most clearly exemplified in obsessional neurotics, and would then familiarize himself with Loukas' theory of intra-psychic conflict, he would get some inkling of the answer he was looking for.

It may be remembered that Loukas was a medical man of wide attainments, who had written a standard work on the rise of the Christian Church. Karina Stephena argued cogently against Joadus that experience showed that intra-psychic conflict was a reality. Being a conflict, it must be between opposing entities: being intra-psychic, these entities must be non-material. Inasmuch as the results of the conflict were shewn in the behaviour of obsessional neurotics, the entities must be evil and not good. Our brief synopsis does not do justice to Karina's triumphant repudiation of the scepticism of Joadus; but this repudiation was widely admitted by leading philosophers of the age to be a valuable contribution to religious psychology.

2. CONCLUDING REMARKS BY THE DEAN OF ST. PAUL'S

The discussion seems to me disappointing, though interesting. It is not really a discussion but a series of comments on a long original text. The reader should bear in mind that no statement of the problem as it appears to those whose approach is radically different from Dr. Waddington's has been given. Obviously a short note is not the vehicle for such a statement, but I think it is relevant to say briefly what the problem, in my opinion, really is. The problem of ethics arises only in the minds of self-conscious reflective persons and it concerns primarily fully voluntary action. A man finds himself pursuing certain general ends, such as pleasure, reputation, social reform, wealth, happiness (as distinct from pleasure), truth, aesthetic satisfaction, artistic creation, the vision of God. The question arises, since some of these ends must sometimes conflict, which if any of these ends is the supreme good or perhaps if some of these apparent goods are really good. Closely connected with this problem is that of the kind of character and conduct which a rational being would approve. Evidently there are differences of opinion and the task of Ethics is to discover some principle by which these differences can be resolved. Now I do not see how this problem can be solved by empirical methods. No amount of observation can answer the question nor can inductive reasoning from the facts about what men do desire determine what they would desire if they were completely rational. I do not say that genetic studies of ethical phenomena have no relevance to the ethical problem. I think Dr. McNeile Dixon somewhere remarks that theories about the origin and development of language have some relevance to the understanding of Shakespeare, though a remote one. Much the same might be said about the relevance of the natural history of ethical development, racial and individual, to the central problem of ethics.

I wish I knew what people meant by the word "evolution". It seems to be highly ambiguous. Sometimes it seems to mean nothing more than a series of more or less continuous changes. Sometimes it means the origin of species by natural selection. Sometimes it means historical development in which human will and thought are the motive forces. Again, sometimes evolution appears to be conceived as a teleological process, sometimes as an undirected one. I really do not know which of these meanings is assumed in the discussion. I am glad that someone has said a kind word for Herbert Spencer, because he had the great virtue of intelligibility. We know what he meant by evolution—

the passage from a condition of "undifferentiated, incoherent homogeneity" to one of "differentiated, coherent heterogeneity". However absurd the conclusions to which Spencer's view leads, when he maintains that good conduct is "more evolved" conduct, we know what he means. The plausibility of the theory, such as it is, lies in the fact that, in some cases, higher values are associated with greater complexity of structure and function. But the value is not the complexity—the complexity is a concomitant and often a very inconvenient one. There is no value in complexity as such.

I must return to the difficulty which I feel about the attempt to extract the moral "ought" from psychological "compulsions". No one has denied that a compulsion is an irresistible and irrational tendency to do or refrain from doing certain things. I think no one will deny that the saner we become the less we are at the mercy of "compulsions", nor the converse that to be the victim of compulsions is a form of insanity. If moral obligation is a species of compulsion, it is a kind of insanity. I maintain, on the contrary, that the perfectly good man would be the perfectly sane man who directed himself by ideals consciously adopted by his reason, in short the very opposite of the man who is directed by impulses from the dark underworld of the psyche.

This has been a good party and all the better perhaps because everyone has talked about what interested him without too much regard for the question at issue. What does it matter, for example, whether some followers have misinterpreted Karl Marx? But there are some expected guests who are absent—the Logical Positivists, who claim to be the only true empiricists. We miss them. I should like to hear them, for example, on the "Super-ego".

Of course there is one view of evolution and development on which it would be at least plausible to hold that evolution is necessarily in the direction of the good, though not that good is equivalent to evolution. A very thorough-going teleology might support such a thesis. If the whole process of existence is directed towards some end which is supremely good in so direct and detailed a manner that every change has its necessary place in the providential plan, then it would be likely that co-operation in producing the next step of the process would always be good, if co-operation in such a theory had any meaning. I do not think that Dr. Waddington holds this view and I suppose he would agree that degeneration is also a fact of experience.

At the risk of appearing pertinaceous I must ask again, what is the ground, on Dr. Waddington's hypothesis, for asserting that there are some things for which a man ought to be willing to die—that it is reasonable to do so? This may seem a harsh intrusion of practical life into an academic discussion, but I believe that it is most pertinent. I do not know what the answer may be. Would you say, "You feel a compulsion?"; but that equates the hero with the lunatic. I call attention to the remarkable passage about the hand in Dr. Waddington's original essay. The statement that it is well not to put your hand into the fire is not based on anything else except the fact that if you do it will cease to be a hand: and existence is its own justification; hands are the kind of things which do not go in fires. Self-destruction of an entity only comes into question when there also exists some large unit of which that entity is a part, and it only occurs when this more inclusive unit is more powerfully energized in the dynamic system of the super-ego. This seems to me a somewhat inadequate reason for holding that in some circumstances, well known alas! to us, it is the duty of men to put not their hands only but their whole bodies into the fire.

3. AN AMERICAN OPINION, BY PROFESSOR CHAUNCEY D. LEAKE

The Relations between Science and Ethics

To the interesting discussion aroused by Dr. C. H. Waddington may be added comments reflecting United States opinion on the matter.

SOME FINAL COMMENTS 131

It was remarkable that three leading American biologists, representing the east, mid-continent, and west, should have come to about the same conclusion at the same time regarding a biological basis for ethics. Different approaches led to the same general position on the part of Professor E. G. Conklin, emeritus professor of biology at Princeton University, Professor C. Judson Herrick, emeritus professor of neurology at the University of Chicago, and Professor Samuel J. Holmes, professor of zoology at the University of California. Conklin[1] says, "Biologically life is maintained by continual balance, co-operation, compromise, and the same principles apply to the life of society. The highest level of human development is attained when purpose and freedom, joined to social emotions, training and habits, shape behaviour not only for personal but also for social satisfactions. Conduct bringing the broader and more lasting satisfactions is the better." According to Herrick,[2] "That social stability upon which the survival and comfort of the individual depend and that moral satisfaction upon which his equanimity, pose and stability of character depend arise from the maintenance of relations with his fellow men which are mutually advantageous." Holmes[3] says, "Morality becomes just one phase of the adjustment of the organism to its conditions of existence. As a good body is one which runs smoothly and efficiently in the maintenance of its vital functions, so a good man is one whose conduct not only maintains his own life on an efficient plane, but conduces to the enhancement of the life of his social group." Both Conklin and Herrick would agree with Holmes in saying, "Peoples may believe that their moral customs derive from a supernatural source, but one potent reason for their adoption is their conduciveness to survival."

These statements suggest that American biologists have come to the same position as Dr. Waddington in regard to the nature of science's contribution to ethics, that is, in

[1] *Scientific Monthly*, 49, 295 (1939).
[2] *Ibid.*, 49, 99 (1939).   [3] *Science*, 90, 117 (1939).

revealing the character and direction of evolution with the elucidation of the consequences "in relation to that direction, of various courses of human action". Our British colleagues may recall Conklin's volume, *The Direction of Human Evolution*, which was published in 1921, and which offers much detailed evidence in support of Dr. Waddington's position.

At the 1940 Christmas meeting of the American Association for the Advancement of Science in Philadelphia, the Section on Historical and Philological Sciences held a symposium on *Science and Ethics*. Participating in this symposium, over which I had the honour of presiding, were Professors Herrick Conklin, Holmes, Teggart, Mackay, Galdston, de Santillana, Sigerist, Sarton, Shryock, Gerard, Birkhoff, and Mayer. At the conclusion of the discussion, the section unanimously agreed to a descriptive statement which seems justifiably inducible from data now available. While taking into account criticisms of the intellectual validity of traditional ethical statements as raised by psychology, anthropology, dialectic materialism, or logical positivism, the statement of these American men of science indicates that they are willing to agree, at our present "level of analysis" as Dr. C. D. Darlington might put it, that certain biological generalities have moral consequences. The recognition by conscious individuals of these consequences, results in "ethical principles as actual psychological compulsions derived from the experience of the nature of society".

The statement may be put in a formal manner: The probability of survival of a relationship between individual humans, or between groups of humans, increases with the extent to which the relationship is mutually satisfying and advantageous. This principle was first formulated in this manner at a memorable seminar in the Santa Cruz redwoods in July 1939, when the Pharmacology Laboratory of the University of California entertained Professors Conklin, Herrick, and Olaf Larsell.[1] It was then appreciated that

[1] *Scientific Monthly*, 53, 133 (1941).

this formulation is merely a special case of the more general biological principle : The probability of survival of individual, groups, or species of living things increases with the degree with which they can and do adjust themselves harmoniously to each other and to their environment.

The ethical significance of this general principle appears in relation to the common biological urges for survival and satisfaction. Consciousness of the operation of this generality suggests the wisdom of such altruistic, considerate, and magnanimous conduct as is intuitively considered 'good' in all ethical systems. The social customs and conventions now with us have so far exhibited survival value in a Darwinian sense. We may apply evolutionary criteria to them and attempt the formulation of a *modus operandi*. Such a formulation constitutes the statement. The principle operates, whether we as humans are conscious of it or not. To promote the conscious appreciation of such natural principles is part of the business of science. There appears to be scientific justification for what philosophers have maintained for centuries, namely, that knowledge of ourselves and of our environment has in itself ethical significance and moral consequence.

CONCLUDING REMARKS BY C. H. WADDINGTON

Yes, I think one can agree with Dr. Matthews that it has been a pleasant party—a bottle party, perhaps, as everyone, even of those who dropped in uninvited, has brought something stimulating with him. When the editor of *Nature* issued his original invitations, I unexpectedly found myself the Lion of the Evening—though a little uncertain whether I was cast as Lion Rampant, expected to turn my back upon the infernal fires and propound moral aphorisms from the centre of the hearth rug, or as Lion Recumbent, whose bones the guests, a selected team of vultures, were invited to amuse themselves by picking. Being of a sanguine temper, I adopted the first pose. And

I shall take the liberty of bringing matters to a close by drawing together the scattered conversations into a short summary of the thesis which, it seems to me, has still escaped conclusive refutation.

1. For every human being, there are some propositions which he considers to be ethical, that is to say, to relate to goodness and badness. The qualities of goodness and badness are recognized as such, and are not identical with any other qualities, such as pleasurableness, desirability, etc.

2. Most discussions of ethics start by attempting to define goodness in terms of other concepts such as those just mentioned. The present thesis starts in quite a different way, by considering the processes which lead to the formation of the concept of the good.

3. Recent psychological work has shown that the concept of the good, and moral ethical feelings in general, are connected with a large system of partly unconscious motives and impulses.[1] These have been referred to (in the early part of the discussion) as the super-ego, although more strictly that name should be reserved for the earlier-formed part of them, while the later-formed part, which has been more influenced by rational thought, should be called the Ego-ideal.

4. Even the earliest formed part of the super-ego, and thus the most primitive notions of the good, are not formed until about the third or fourth month of post-natal life. They are formed by the interaction of the strivings of the child (motivated by pleasure-pain feelings and physiological drives) with his surroundings.

5. When we speak of the concept of the good being formed at this time, this means exactly what it says. The baby is

---

[1] Dr. Matthews has I admit, put his finger on a stupid mistake when he takes me to task for speaking of ethical feelings as *compulsions*. A compulsion must, I suppose, always succeed in controlling behaviour. This I never meant to imply, as I should have thought the subsequent discussion made clear. I should have used the word "impulsion", since all I intended to emphasize was that ethical feelings are not mere emotions, such as the feeling of pleasure or happiness, but are drives or tendencies towards certain courses of action.

not merely learning how to attain certain ends which it already has; it is acquiring the realization of what its ends are. And it is reaching this realization through the observation of the results of its behaviour. There is therefore ultimately no distinction between ends and means; the ends are formulated (if such a word can be used of the crude, unintellectual process concerned) as a result of trying out various means. An ethical system is in fact as much an adaptation to the environment as a theory of chemistry.

6. Only certain of the propositions which the child forms as the result of its experience become associated with ethical feelings. These ethical propositions concern primarily the child's relation with other people, particularly its father and mother, and are thus social in character. Ethical propositions in fact, fundamentally deal with the conditions for social existence.

7. It is an observed fact, that the ethical beliefs arising within a society are not such as merely to ensure the existence of that society; they normally tend to produce changes of the social order. Now almost the only steps which have recently occurred in the evolution of mankind have been changes in social order; and it is mainly such changes which can be looked for in the evolution of the foreseeable future. The formation of an ethical system is thus a mechanism which both enables man to live in society and provides the motivation for attempts at further evolutionary advance.

8. It is not, of course, pretended that the mechanism by which ethical systems are formed is sufficiently automatic to ensure that everyone forms the same system of beliefs. But once it is recognized that the biological function of ethical systems is to act as the machinery on which the evolution of society depends, it becomes possible to define the ethical system which has a general validity for mankind as a whole; it must be that ethical system which has actually been effective in guiding the evolution of man as a whole, and its characteristics can therefore be deduced from the nature of the evolutionary changes which it has produced.

9. The social evolution of man appears to be a carrying-forward of the main features of animal evolution (e.g. in increasing control over the environment, capacity to react, to the relations between objects, etc.). We would therefore probably be justified in deducing our general ethical system from the characteristics of evolution as a whole rather than merely from human evolution. But it would in practice not be necessary to do this unless mankind was threatened with a regression to a level below that at which human life transcends that of the animals.

10. The argument in paragraph 8 states that although the nature of a generally valid ethical system is not directly known, it can be deduced because that system is concerned in the production of a process (human evolution), the course of which is open to examination. This argument clearly assumes that the course of evolution has not been influenced by some other agency which has been more powerful than the ethical strivings of mankind. There seems no reason to doubt this assumption under present conditions, but it would clearly become invalid if there was an overwhelming change in environmental conditions, such as would be caused by the cooling down of the sun, etc. If a general evolutionary regression took place under such conditions it seems that there would be no way of determining a generally valid ethical system applicable under the circumstances.

11. It is then clear that science is in a position to make a contribution to ethics, since ethical systems are derived from the observation of facts of the kind with which science deals. And the nature of science's contribution is also clear; it is the revelation of the character and of the direction of the evolutionary process in the world as a whole, and the elucidation of the consequences, in relation to that direction, of various courses of human action.

Perhaps it would be impolite merely to pass over in silence Dr. Matthews' so kind invitation to enter his study and share the melancholy fate of Herbert Spencer. Poor Spencer! He was cajoled or bamboozled by literary men into behaving as though he was talking not about phenomena but about forms of words. Rather half-heartedly he stated, or implied, that what he meant by "good" was "productive of pleasure"; and the critics (e.g. Moore) showed their gratitude at not being asked to raise their eyes from their books by pointing out that that would not do at all. Similarly he made the tactical error of suggesting that evolution can be adequately summarised in a single phrase, such as "increase in complexity"; and the *literati* indulged in a hearty laugh at the "absurdity of the conclusions to which Spencer's view leads."

I have at least learned that lesson, and am not so easily caught by that lure. By "evolution" I refer to a process which I can point out to you if you come with me to a good biological museum. If you insist on substituting for the actual phenomena a collection of words, I might be passably content with a shelf-ful of treatises, but I should insist that even they are a description which may or may not be adequate. Again, by "the individual's idea of the good" I mean a set of actual beliefs, feelings and motives, and I think I can indicate to you which set, if there is any doubt. On the other hand, "the generally valid ethical system" is a logical abstraction which one is attempting to derive from these observed phenomena. It should be possible to define it in terms of the observables from which it is constructed; and this is what I have tried to do.

This book has been concerned almost entirely with the preliminary task of establishing the validity of this point of view as to the nature of ethics, and it has left on one side the discussion of what the thesis, if it were adopted, would imply about particular ethical problems. These implications are, however, by no means trivial. For instance, if ethical systems are admitted to be important variables in the evolutionary process in man, the existence of evil ethical systems becomes comparable to the existence of deleterious genotypes. We know that, on the level of biological evolution, progress is only possible if hereditary variations occur. The process of mutation ensures that variations in all possible directions shall be constantly available in a large population; but the great majority of these are harmful. All evolving species have to pay some price, in a certain lack of adaptation, in order to retain the potentiality for further evolutionary change. It may not be implausible to suggest that on the social level of evolution to which man has attained, there is also an inevitable price, in the existence of evil, to be paid to keep open the path to progress. Dr. Matthews' perfectly sane and perfectly good man would be in an evolutionary cul-de-sac, a position which, at any rate on the biological level, seems to be extremely difficult to maintain. But in the world of to-day, the formation of ethical ideas or of the Ego-ideal, as Dr. Stephen wished to call it, proceeds by means of processes of projection and introjection which seem perhaps only too well calculated to produce abundant variability.

Again the moral problem of death, which Dr. Matthews raised, would be brought by this point of view into relation with its known biological functions. "What is the ground, on Dr. Waddington's hypothesis", he asks, "for asserting that there are some things for which a man ought to be willing to die—that it is reasonable to do so? . . . Would you say, 'You feel a compulsion'? but that equates the hero with the lunatic." I should not, in dealing with this question, forget the classical remark of Friar John to Pantagruel on the subject: "Die? We all must." The theory of biological evolution has for long regarded an individual as a particular experiment in the formation of genotypes; the death of the individual, coupled with the mechanism of heredity, is a necessary part of the process of evolution. If the Mesozoic reptiles had been immortal, man could never have appeared on the scenes. In so far as ethics are taken to be involved in evolution, this argument can be carried over from the biological field to the moral; the sacrifice of personality which the individual makes at death can be seen as a contribution to the welfare of the human race.

In particular circumstances an individual may be called upon to make this sacrifice by a conscious effort. If I substitute the word "impulsion" for "compulsion" have I not gone far enough to meet Dr. Matthews' criticism of the psychological mechanism I suggested? A man who lays down his life must surely feel a conflict of motives; not only a drive to serve the values of self, but also one towards the values of something which will usually be more inclusive than self (the morphia addict who consciously drugs himself to death would be an exception in that the "death motives" would be less inclusive).

Dr. Matthews has expressed the wish to have heard representatives of several other points of view. This matter of the conflict of motives is the context in which I should have most liked a contribution from another standpoint. This book has dealt almost entirely with what might be called macro-ethics, the values concerned in the progress of the human race as a whole. One remembers that rugged idealist William Blake: "General Good is the plea of the scoundrel, hypocrite and flatterer. . . . To generalize is to be an idiot. . . . He who would do good to another must do it in Minute Particulars." The human race is a collection of individuals; it has no collective consciousness. Can the General Good justify the destruction of individual values? A point of view which, while not perhaps finally answering this question in the negative, expresses the difficulty very keenly, is the theme of William Bowyer's recent and important Autobiography, *Brought Out in Evidence*; I quote a few sentences:

It must be clear to all not blinded by habitual acceptance or afraid to face an unpleasant fact that every living creature maintains its life by consuming others. . . . The happy bird of romantic legend will if more closely observed be seen glancing frantically about in terror of a host of enemies, desperate for food, as it tears to pieces the living worm it has wrenched from the ground. Its life no less than its death is as much a hideous storm of terror as that of the others. . . . For a creature, born without its consent, finally to die of "old age", worn out as a machine is worn out, thrown aside as if forgotten, after being tantalized with a fugitive happiness, played with as a cat plays with a mouse, could appear only as a wanton and cruel mockery if it were not accepted unthinkingly as familiar and "natural".

The scientist would probably question, as unduly anthropomorphic, the assertion that the bird lives in a continuous state of terror and desperation; but Bowyer's is a poetic statement, and a powerful and explicit one, of a real dilemma of feeling. The progress of social evolution has in the past, and probably will in the future, led to an increasing value being put on the individual personality. To a greater or lesser extent, this must conflict with the supra-individual values of the human race as whole. It is not easy to see that the conflict will lessen; in fact as civilized self-consciousness deepens, it seems that the struggle between the individual and the social loyalties must become ever more crucial. Perhaps, however, the reference to self-consciousness is at the same time an indication of the nature of a possible synthesis of these dialectical opposites. For with increasing self-consciousness should go increasing intellectual comprehension of the relation of the individual to mankind. The moral conflict of One and Many will not disappear, but reason may be able to focus its dynamism into a single constructive stream.

THE CONTRIBUTORS

The Right Reverend E. W. BARNES, F.R.S., Bishop of Birmingham; born 1874, published several works on pure mathematics, theology, sermons, *Scientific Theory and Religion* (Gifford Lectures, 1933).

J. D. BERNAL, F.R.S., Professor of Physics, Birkbeck College, London; born 1901, published many papers on crystallography and X-ray analysis, also *The Social Function of Science*, an acknowledged leader among the younger scientists, whose wide interests embrace both physical and biological topics.

G. BURNISTON BROWN, Lecturer in Physics, University College, London.

C. D. DARLINGTON, F.R.S., Director of the John Innes Horticultural Institution; born 1903, published many papers on cytology and genetics, *Recent Advances in Cytology*, *The Evolution of Genetic Systems*, etc.

W. G. DE BURGH, F.B.A., Emeritius Professor of Philosophy, Reading University; born 1866, published *The Legacy of the Ancient World*, *From Morality to Religion* (Gifford Lectures, 1938), etc.

H. DINGLE, Professor of Natural Philosophy, Imperial College, London; born 1890, published many papers and books on astronomy and the theory of relativity, also *Through Science to Philosophy*.

H. J. FLEURE, F.R.S., Professor of Geography, Manchester University; born 1877, published papers and books on anthropology and antiquities.

J. B. S. HALDANE, F.R.S., Professor of Biometry, University College; born 1892, published many scientific works on biochemistry and applications of mathematics to biology, also popular essays, *Science and Ethics* (1928), *Heredity and Politics*, *The Marxist Philosophy and the Sciences*, etc.

J. S. HUXLEY, F.R.S., Secretary, The Zoological Society of London; born 1887, published many papers on biology (development, animal behaviour, evolution, etc.), popular essays, *What Dare I Think? The Uniqueness of Man*, etc.

C. E. M. JOAD, Head of Department of Philosophy and Psychology, Birkbeck College, London; born 1891, published many popular essays on philosophy.

M. KLEIN, author of numerous papers on psychoanalysis, with particular reference to children.

C. D. LEAKE, Professor of Pharmacology, University of California Medical School.

The Very Reverend W. R. MATTHEWS, K.C.V.O., Dean of St. Paul's; born 1881, published many theological works.

J. NEEDHAM, F.R.S., Reader in Biochemistry, Cambridge University; born 1900, published many scientific papers, *The Great Amphibium*, *Order and Life*, etc.

A. D. RITCHIE, Professor of Philosophy, Manchester University; born 1891, formerly Lecturer in physiology, published scientific works on that subject, also philosophical books, *The Natural History of Mind*, etc.

M. ROTHSCHILD, independent biologist.

L. S. STEBBING, Professor of Philosophy, Bedford College, London; born 1885, published philosophical articles and books, *Thinking to Some Purpose*, *Ideals and Illusions*.

K. STEPHEN, psychoanalyst, published *The Misuse of the Mind*, etc.

C. H. WADDINGTON, Lecturer in Zoology, Cambridge University; born 1905, published papers and books on biology (development and genetics), also *The Scientific Attitude*.

INDEX OF NAMES